JOHN DEERE

SHOP MANUAL

More information available at haynes.com
Phone: 805-498-6703

Haynes Group Limited
Haynes North America, Inc.

ISBN 10: 0-87288-471-6
ISBN-13: 978-0-87288-471-7

Disclaimer

There are risks associated with automotive repairs. The ability to make repairs depends on the individual's skill, experience and proper tools. Individuals should act with due care and acknowledge and assume the risk of performing automotive repairs.

The purpose of this manual is to provide comprehensive, useful and accessible automotive repair information, to help you get the best value from your vehicle. However, this manual is not a substitute for a professional certified technician or mechanic.

This repair manual is produced by a third party and is not associated with an individual vehicle manufacturer. If there is any doubt or discrepancy between this manual and the owner's manual or the factory service manual, please refer to the factory service manual or seek assistance from a professional certified technician or mechanic.

Even though we have prepared this manual with extreme care and every attempt is made to ensure that the information in this manual is correct, neither the publisher nor the author can accept responsibility for loss, damage or injury caused by any errors in, or omissions from, the information given.

JD-58, 9S1, 14-160

Information and Instructions

This individual Shop Manual is one unit of a series on agricultural wheel type tractors. Contained in it are the necessary specifications and the brief but terse procedural data needed by a mechanic when repairing a tractor on which he has had no previous actual experience.

The material is arranged in a systematic order beginning with an index which is followed immediately by a Table of Condensed Service Specifications. These specifications include dimensions, fits, cleararances, capacities and tune-up information. Next in order of arrangement is the procedures section.

In the procedures section, the order of presentation starts with the front axle system and steering and proceeds toward the rear axle. The last portion of the procedures section is devoted to the power take-off and power lift systems.

Interspersed where needed in this section are additional tabular specifications pertaining to wear limits, torquing, etc.

How to use the index

Suppose you wnt to know the procedure for R&R (remove and reinstall) of the engine camshaft. Your first step is to look in the index under the main heading of "Engine" until you find the entry "Camshaft." Now read to the right. Under the column covering the tractor you are repairing, you will find a number which indicates the beginning paragraph pertaining to the camshaft. To locate this paragraph in the manual, turn the pages until the running index appearing on the top outside corner of each page contains the number you are seeking. In this paragraph you will find the information concerning the removal of the camshaft.

Common spark plug conditions

NORMAL
Symptoms: Brown to grayish-tan color and slight electrode wear. Correct heat range for engine and operating conditions.
Recommendation: When new spark plugs are installed, replace with plugs of the same heat range.

WORN
Symptoms: Rounded electrodes with a small amount of deposits on the firing end. Normal color. Causes hard starting in damp or cold weather and poor fuel economy.
Recommendation: Plugs have been left in the engine too long. Replace with new plugs of the same heat range. Follow the recommended maintenance schedule.

TOO HOT
Symptoms: Blistered, white insulator, eroded electrode and absence of deposits. Results in shortened plug life.
Recommendation: Check for the correct plug heat range, over-advanced ignition timing, lean fuel mixture, intake manifold vacuum leaks, sticking valves and insufficient engine cooling.

CARBON DEPOSITS
Symptoms: Dry sooty deposits indicate a rich mixture or weak ignition. Causes misfiring, hard starting and hesitation.
Recommendation: Make sure the plug has the correct heat range. Check for a clogged air filter or problem in the fuel system or engine management system. Also check for ignition system problems.

PREIGNITION
Symptoms: Melted electrodes. Insulators are white, but may be dirty due to misfiring or flying debris in the combustion chamber. Can lead to engine damage.
Recommendation: Check for the correct plug heat range, over-advanced ignition timing, lean fuel mixture, insufficient engine cooling and lack of lubrication.

ASH DEPOSITS
Symptoms: Light brown deposits encrusted on the side or center electrodes or both. Derived from oil and/or fuel additives. Excessive amounts may mask the spark, causing misfiring and hesitation during acceleration.
Recommendation: If excessive deposits accumulate over a short time or low mileage, install new valve guide seals to prevent seepage of oil into the combustion chambers. Also try changing gasoline brands.

HIGH SPEED GLAZING
Symptoms: Insulator has yellowish, glazed appearance. Indicates that combustion chamber temperatures have risen suddenly during hard acceleration. Normal deposits melt to form a conductive coating. Causes misfiring at high speeds.
Recommendation: Install new plugs. Consider using a colder plug if driving habits warrant.

OIL DEPOSITS
Symptoms: Oily coating caused by poor oil control. Oil is leaking past worn valve guides or piston rings into the combustion chamber. Causes hard starting, misfiring and hesitation.
Recommendation: Correct the mechanical condition with necessary repairs and install new plugs.

DETONATION
Symptoms: Insulators may be cracked or chipped. Improper gap setting techniques can also result in a fractured insulator tip. Can lead to piston damage.
Recommendation: Make sure the fuel anti-knock values meet engine requirements. Use care when setting the gaps on new plugs. Avoid lugging the engine.

GAP BRIDGING
Symptoms: Combustion deposits lodge between the electrodes. Heavy deposits accumulate and bridge the electrode gap. The plug ceases to fire, resulting in a dead cylinder.
Recommendation: Locate the faulty plug and remove the deposits from between the electrodes.

MECHANICAL DAMAGE
Symptoms: May be caused by a foreign object in the combustion chamber or the piston striking an incorrect reach (too long) plug. Causes a dead cylinder and could result in piston damage.
Recommendation: Repair the mechanical damage. Remove the foreign object from the engine and/or install the correct reach plug.

SHOP MANUAL

JOHN DEERE

SERIES

2150—2155—2255—2350—2355—2355N—2550—2555

Tractor serial number is stamped in right side of front support and is also located on a plate attached to right side of front support. Engine serial number is stamped on a plate on right side of engine cylinder block.

INDEX (By Starting Paragraph)

DUAL DIMENSIONS

This shop manual provides specifications in both the Metric (SI) and U.S. Customary systems of measurement. The first specification is given in the measuring system used during manufacture, while the second specification (given in parenthesis) is the converted measurement. For instance, a specification of "0.28 mm (0.011 inch)" would indicate that the equipment was manufactured using the metric system of measurement and the U.S. equivalent of 0.28 mm is 0.011 inch.

CONDENSED SERVICE DATA

	2150 2155 2255	2350 2355	2355N	2550	2555 Collar Shift Transmission	2555 Synchronized Transmission
GENERAL						
Engine Model	3179DL	4239DL	3179TL	4239DL	4239DL	4239TL
No. of Cylinders	3	4	3	4	4	4
Bore			106.5 mm (4.19 in.)			
Stroke			110 mm (4.33 in.)			
Displacement	2.94 L (179 cid.)	3.92 L (239 cid.)	2.94 L (179 cid.)	3.92 L (239 cid.)	3.92 L (239 cid.)	3.92 L (239 cid.)
Compression Ratio** . .			17.8:1			
Battery Terminal Grounded			Negative			
TUNE-UP						
Firing Order	1-2-3	1-3-4-2	1-2-3	1-3-4-2	1-3-4-2	1-3-4-2
Valve Clearance — Inlet			0.35 mm (0.014 in.)			
Exhaust			0.45 mm (0.018 in.)			
Engine Low Idle—Rpm .			Refer to paragraph 88			
Engine High Idle—Rpm			Refer to paragraph 88			
Pto Power at Rpm	*1	*2	*3	*4	*5	*5

*1 2150 - 34 kW (45 Hp) @ 2500 rpm for models before engine SN 571079 CD;
   37 kW (50 Hp) @ 2500 rpm for models after engine SN 571078 CD.
 2155 - 34 kW (45 Hp) @ 2300 rpm for synchronized transmission;
   34 kW (45 Hp) @ 2500 rpm for collar shift transmission.
 2255 - 34 kW (45 Hp) @ 2500 rpm for models before engine SN 571079 CD;
   37 kW (50 Hp) @ 2500 rpm for models after engine SN 571078 CD.

*2 2350 - 41 kW (55 Hp) @ 2550 rpm
 2355 - 41 kW (55 Hp) @ 2300 rpm for synchronized transmission;
   41 kW (55 Hp) @ 2500 rpm for collar shift transmission.

*3 2355N - 41 kW (55 Hp) @ 2300 rpm for synchronized transmission;
   41 kW (55 Hp) @ 2500 rpm for collar shift transmission.

*4 2550 - 48 kW (65 Hp) @ 2550 rpm

*5 2555 - 48 kW (65 Hp) @ 2300 rpm for synchronized transmission;
   48 kW (65 Hp) @ 2500 rpm for collar shift transmission.

** Compression ratio of some early models may be 16.8-17.4 : 1.

	2150 2155 2255	2350 2355	2355N	2550	2555 Collar Shift Transmission	2555 Synchronized Transmission

SIZES—CLEARANCES

Crankshaft Main
  Journal Diameter . . .   79.34-79.36 mm
       (3.123-3.124 in.)

Crankpin Diameter . . .   69.80-69.82 mm
       (2.748-2.749 in.)

Piston Pin Diameter —
  Small pin . . . .   34.92-34.93 mm
       (1.374-1.375 in.)
  Large pin . . . . . .   41.27-41.28 mm
       (1.624-1.625 in.)

Main Bearing Clearance   0.03-0.10 mm
       (0.001-0.004 in.)

Rod Bearing Clearance .   0.03-0.10 mm
       (0.0012-0.004 in.)

Camshaft Journal Clearance   0.10-0.15 mm
       (0.004-0.006 in.)

Crankshaft End Play —
  W/ 2-Piece Thrust Bearing   0.05-0.20 mm
       (0.002-0.008 in.)
  W/ 5-Piece Thrust Bearing   0.025-0.43 mm
       (0.001-0.017 in.)
  All Dubuque Built Engines ***   0.025-0.33 mm
       (0.001-0.013 in.)

Camshaft End Play . . .   0.05-0.23 mm
       (0.002-0.009 in.)

Piston Skirt Clearance —
  All Naturally
    Aspirated Engines . .   0.08-0.14 mm
       (0.003-0.005 in.)

  Turbocharged Engines —
    Dubuque Built Engines ***   0.08-0.15 mm
       (0.003-0.006 in.)

  Engines Built At
    Other Locations
      Before CD600000   0.14-0.20 mm
       (0.005-0.008 in.)
      After CD599999   0.08-0.15 mm
       (0.003-0.006 in.)

*** First 2 characters of engine Serial Number will be "TO."

CAPACITIES
All capacities are approximate and may be different than shown.

	2150 2155 2255	2350 2355	2355N	2550	2555 Collar Shift Transmission	2555 Synchronized Transmission
Cooling System			Refer to paragraph 89			
Crankcase (with filter) .	6.5-7.0 L (6.9-7.4 qt.)	8.5 L (9 qt.)	6.5 L (6.9 qt.)	8.5 L (9.2 qt.)	10.5 L (11.2 qt.)	10.5 L (11.2 qt.)
Fuel Tank	74 L **** (19.6 gal.)	91 L (24 gal.)	74 L **** (19.6 gal.)	91 L (24 gal.)	84 L (22.2 gal.)	84 L (22.2 gal.)

CONDENSED SERVICE DATA (CONT.)

	2150 2155 2255	2350 2355	2355N	2550	2555 Collar Shift Transmission	2555 Synchronized Transmission
CAPACITIES (Cont.)						
Transmission & Hydraulic —						
Synchronized	50-57 L (14-15 gal.)	59 L (15.8 gal.)	50-53 L (13-14 gal.)	50 L (13.2 gal.)	50-55 L (13.2-14.5 gal.)	50-55 L (13.2-14.5 gal.)
Collar Shift	36-42 L (9.5-11 gal.)	42.0 L (11.1 gal.)	36-42 L (9.5-11 gal.)	33 L (8.7 gal.)	47.5-52.5 L (12.5-13.9 gal.)	47.5-52.5 L (12.5-13.9 gal.)
Front-Wheel Drive —						
Axle Housing	5.3 L (5.6 qt.)	5.0 L (5.2 qt.)	3.5 L (3.7 qt.)	5.0 L (5.2 qt.)	5.3 L (5.6 qt.)	5.3 L (5.6 qt.)
Planetary (each)	0.75 L (0.8 qt.)	0.75 L (0.8 qt.)	0.5 L (0.52 qt.)	0.75 L (0.8 qt.)	0.75 L (0.8 qt.)	0.75 L (0.8 qt.)

**** 2155 models and 2355N models after SN 621 999 fuel tank capacity is 67 L (17.7 gals.).

FRONT SYSTEM
(TWO-WHEEL DRIVE)

AXLE AND SUPPORT

All Models

1. AXLE CENTER MEMBER. To remove front axle assembly, attach a hoist that is suitable for raising the tractor and the front support in such a way that it will not interfere with the removal of the axle. Disconnect drag link from the bellcrank (13—Fig. 1 or Fig. 1A) on models so equipped. On models with hydrostatic steering, disconnect hoses from the steering cylinders and cover openings to prevent the entry of dirt. On all models, support the axle with a suitable jack or special safety stand to prevent tipping while permitting the axle to be moved safely. Remove the slotted nut (1), special pivot bolt (6), washers and shims (3). Move axle rearward until free from pivot tube (4) and pin (7—Fig. 2). Raise the front of tractor with previously attached hoist and carefully roll axle away from under front support. The bellcrank (13—Figs. 1 or 1A) is supported in needle bearings of some (fixed tread) models and in tapered roller bearings of other (adjustable tread axle) models. Both types of bearings should be lubricated by packing with EP multi-purpose grease. The end play of bellcrank supported in needle roller bearings (12—Fig. 1) should be adjusted to 0.1 mm (0.004 inch) or less by adding the required number of shims (11). The bellcrank (13—Fig. 1A) supported in tapered roller bearings should be correctly located vertically by adding the required number of shims (11) between the bearing cup of upper bearing and bore in axle center member. Correct selection of shims is determined by measuring the depth of bearing bore, then subtracting 18.4 mm (0.724 inch). The result is thickness of shims (11) required. Shim thickness can be rounded to the nearest tenth (0.1) mm when assembling. Tighten the special nut (22) to a torque of 55 N·m (40 ft.-lbs.) on 2150 and 2255 models; 60 N·m (45 ft.-lbs.) on other models. On all models, check bushing (8—Fig. 1 or Fig. 1A), front pivot tube (4), rear pivot pin (7—Fig. 2) and bushing (24) for excessive wear and renew if necessary. Be sure to align passage in bushing with appropriate grease fitting if new bushing is installed. Crossed grooves in bushing (8—Fig. 1 or Fig. 2) should be down and slot in bushing (24—Fig. 2)

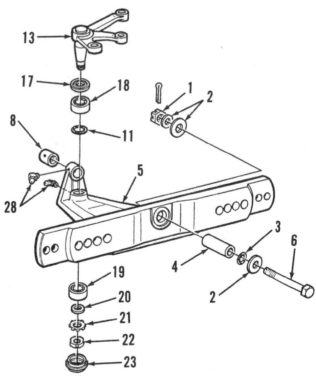

Fig. 1A—Front axle center member typical of type used on some Two-Wheel Drive models.

1. Slotted nut	
2. Washer	
3. Shim	17. Sealing disk
4. Pivot tube	18. Upper bearing
5. Axle center member	19. Lower bearing
6. Pivot bolt	20. Washer
8. Bushing	21. Lock washer
11. Shim	22. Nut
13. Steering bellcrank	23. Cap
	28. Grease fittings

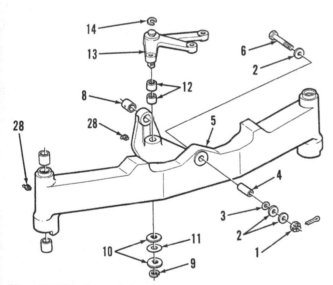

Fig. 1—Fixed tread front axle typical of type used on some Two-Wheel Drive models.

1. Slotted nut	9. Snap ring
2. Washer	10. Washers
3. Shim	11. Shim
4. Pivot tube	12. Needle bearings
5. Axle	13. Steering bellcrank
6. Pivot bolt	14. Snap ring
8. Bushing	28. Grease fittings

6

should be up. Reverse removal procedure when assembling. Tighten slotted nut (1—Fig. 1 or Fig. 1A) to 300 N·m (220 ft.-lbs) torque. End clearance of axle center member should be 0-0.4 mm (0-0.016 inch) and is adjusted by adding or removing shims (3). If cotter pin cannot be installed through slotted nut (1) after tightening to specified torque, tighten enough to align hole with next slot, then install cotter pin.

2. FRONT WHEEL BEARINGS. To remove front wheel hub and bearings, raise and support the front axle extension, then unbolt and remove the tire and wheel assembly. Remove the cap (12—Fig. 3), cotter pin (9), slotted nut (10), washer (8) and outer bearing cone (7). Slide the hub assembly from spindle axle shaft, then remove seal (1) and inner bearing cone (3) from hub. Remove and install new collar (2) if scored or otherwise damaged. Hub is slotted to facilitate removal of bearing cups (4 and 6). Seal (1) used with collar (2) should be driven onto axle with numbered side of seal out toward driving tool. Fill space between seal lips of all seals with EP multi-purpose grease and pack bearings liberally with wheel bearing grease. Reassemble by reversing disassembly procedure. Tighten slotted nut (10) to a torque of 50 N·m (35 ft.-lbs.) while rotating hub. Back nut off to nearest slot and install cotter pin (9), then install cap (12).

3. SPINDLES AND BUSHINGS. To remove spindle (1—Fig. 4 or Fig. 5), first remove the wheel and hub as outlined in paragraph 2. Disconnect tie rod end from steering arm (9C). On models with steering cylinders attached to steering arms, detach the steering cylinder from the steering arm. On all models,

remove cap screw and washers from top of spindle. Remove steering arm (9), shim (2) and upper seal (8), then lower spindle out of axle extension. Models with clamping type steering arm (9—Fig. 4) are not equipped with shim. On all models, remove thrust bearing (5—Fig. 4 or Fig. 5), lower seal (4) and washer (3) from spindle. Clean and inspect parts for wear or other damage and renew as necessary.

Install lower bushing (6) until flush with bottom of bore. Upper bushing (6) should be installed 8.5 mm (0.33 inch) below top of bore on 2155, 2355, 2355N and 2555 models with fixed tread axle, 11 mm (0.430 inch) below top of bore for 2355 and 2555 models with heavy duty adjustable axle. Install upper bushings flush with counterbore of all other types of axles. Bushings are presized and should require no final sizing if carefully installed. Inside diameter of new installed bushings should be 34.94-35.04 mm (1.375-1.379 inches) for 2150 and 2255 models and 38.11-38.21 mm (1.500-1.504 inches) for 2350 and 2550 models.

When reassembling, install thrust bearing (5) on spindle so that numbered side of bearing is facing upward. Add or remove shims (2) to adjust spindle

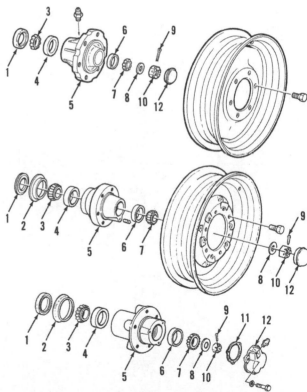

Fig. 3—Exploded view of three front wheel hubs used on Two-Wheel-Drive models. Seal collar (2) is used for seals that are pressed on spindle hub.

1. Oil seal	7. Bearing cone
2. Collar	8. Washer
3. Bearing cone	9. Cotter pin
4. Bearing cup	10. Slotted nut
5. Hub	11. Gasket
6. Bearing cup	12. Cap

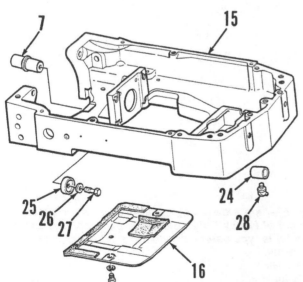

Fig. 2—View of a typical front support casting and associated parts.

7.	Rear pivot pin	25.	Spacer
15.	Front support	26.	Washer
16.	Bottom cover	27.	Cap screw
24.	Front pivot bushing	28.	Grease fitting

end play on models with cap screw retaining steering arm to spindle. On models with steering arm (9C—Fig. 4) clamped to spindle, relocate steering arm to adjust end play of spindle. Spindle end play should be 0.13-1.15 mm (0.005-0.045 inch) for 2355 GP models, 0.76 mm (0.030 inch) for all other models. Tighten steering arm retaining clamping screw (22) to a torque of 120 N.m (85 ft.-lbs.) for all models so equipped. Tighten the retaining cap screw (21) to a torque of 230 N.m (170 ft.-lbs.) for all models except 2150. Tighten the retaining screw (21) on 2150 models so equipped to 120 N.m (85 ft.-lbs.) torque. Balance of reassembly is the reverse of disassembly.

4. TIE RODS AND TOE-IN. Some models are equipped with two tie rods extending from left and right steering arms to steering bellcrank (13—Fig. 1 or Fig. 1A). Other models are equipped with one tie rod extending between left and right steering arms.

Recommended toe-in is 3.0-9.0 mm (1/8-3/8 inch) for 2155 and 2355N models with a fixed tread axle, 3.5-9.5 mm (5/32-3/8 inch) for 2355 and 2555 models with fixed tread axle and 3-6 mm (1/8-1/4 inch) for all other models. To adjust toe-in on models with tie rod ends as shown in Fig. 4, loosen or remove bolts in outer clamps of both tie rods and loosen bolts on inner clamps. When adjusting models with two tie rods, adjust each side an equal amount to obtain proper toe-in. Install and tighten clamp bolts. To adjust toe-in on models with steering cylinders attached to the steering arms (9—Fig. 5), loosen bolt in clamp (13), remove snap ring (16) and withdraw pin (15). Turn tie rod end (12) in or out of tie rod tube (11) as required to obtain proper toe-in. Connect tie rod to steering arm and tighten clamp bolt.

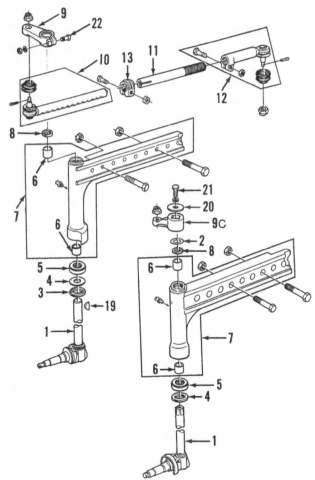

Fig. 4—Axle extension, spindle and associated parts used on some models. Refer also to Fig. 5.

1. Spindle	9. Steering arm (Clamp type)
2. Shim	9C. Steering arm (Splined)
3. Washer	10. Tie rod end
4. Lower seal	11. Tube
5. Thrust bearing	12. Tie rod end
6. Bushings	20. Washer
7. Axle extension	21. Cap screw
8. Upper seal	22. Clamp screw

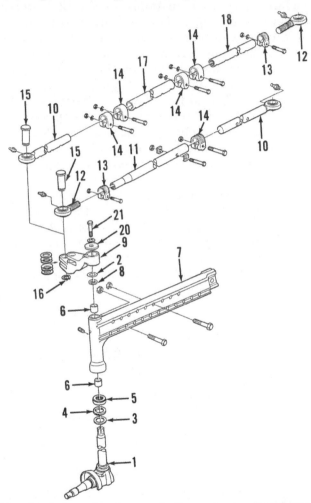

Fig. 5—Axle extension, spindle and associated parts typical of Heavy Duty adjustable axle models. Steering arm (9) is type used with steering cylinders attached to the arms.

1. Spindle	10. Tie rod end
2. Shim	11. Tube
3. Washer	12. Tie rod end
4. Lower seal	13. Clamp
5. Thrust bearing	14. Clamp
6. Bushings	15. Pin
7. Axle extension	16. Snap ring
8. Upper seal	17. Outer tube
9. Steering arm	18. Inner tube

FRONT-WHEEL DRIVE SYSTEM
(2150 MODELS SO EQUIPPED)

A mechanical front-wheel drive is used on 2150 models and is controlled by an electric solenoid/hydraulic valve that operates a multiple disc clutch. This clutch is located in the bottom front of the synchronized transmission case. A drive shaft with two "U" joints connects the clutch unit to front axle. Front axle is equipped with a self-locking differential and a planetary drive at each of the front wheel hubs. The gears that drive the clutch are integral with the synchronized transmission.

> **CAUTION: When servicing a front-wheel drive equipped tractor with rear wheels supported off ground, engine running and transmission in gear, always raise front wheels off ground too. Loss of electrical power or transmission hydraulic system pressure will engage front driving wheels. Rear wheels will be pulled off support if front wheels are not raised.**

TROUBLE-SHOOTING

5. Some problems that may occur with front-wheel drive and their possible causes are as follows:

1. Front-wheel drive not engaging. Could be caused by:
   a. Dash electrical switch defective. Check to be sure that electrical circuit is open when key switch and dash "MFWD" switch are both ON. (The engine does not need to be running when making this check.) Outside of control solenoid should not have strong magnetic field as checked by placing a screwdriver near solenoid.
   b. Solenoid/hydraulic valve spool sticking open. The valve may stick if the valve attaching screws are improperly or unevenly tightened.
   c. Disc clutch pack worn. To check for slippage with engine stopped, set parking brake, slide the drive shaft shield forward and remove the rear guard mounting bracket. Raise the left front wheel off ground using a floor jack, then use a pipe wrench or other suitable tool attached to the clutch output shaft to measure torque required to cause clutch to slip. Clutch should not slip at less than 800 N·m (600 ft.-lbs.) torque. Torque can be measured using a pipe wrench and suitable handle extending pipe by dividing your total body weight into the desired torque and the result will be the distance from shaft center line to apply your weight. An example would be 600 ft.-lbs. (desired torque) ÷ 180 lbs. (your body weight) = 3.33 ft. (distance from shaft center line). To convert the distance to inches, multiply 3.33 × 12 = 39.96 inches.

   d. Mechanical failure in front-wheel drive.

2. Front-wheel drive not disengaging. Can be checked by shifting transmission to "Park," then jacking up left front wheel. With key switch ON, dash "MFWD" switch OFF, engine operating at 1000 rpm and oil temperature warmed to at least 38° C (100° F), it should be possible to rotate the left front wheel by hand. Failure to disengage could be caused by:
   a. Electrical fault (switch, fuse or wiring). Check for magnetic field at solenoid valve using a screwdriver or similar object near the valve with key switch ON and dash "MFWD" switch OFF.
   b. Defective solenoid/hydraulic valve (21E or 21L—Fig. 19). The valve may stick if the valve attaching screws (12 or 27) are improperly or unevenly tightened.
   c. Hydraulic pressure too low or leaking piston seal rings. Pressure can be checked as outlined in paragraph 14.

3. Excessive front tire wear. Could be caused by:
   a. Toe-in incorrect.
   b. Front-rear tire combination not as specified.
   c. Incorrect tire inflation.

TIE ROD AND TOE-IN

6. Tie rod ends are not adjustable for wear and faulty units must be renewed.

To check toe-in, first turn steering wheel so that front wheels are in straight ahead position. Measure distance at front and rear of front wheels from rim flange to rim flange at hub height. Toe-in should be 0-3.0 mm (0-0.118 inch). If necessary to adjust toe-in, disconnect tie rod end from steering knuckle arm. Loosen clamp bolt and adjust tie rod end in or out as required. Reinstall tie rod end to steering arm and tighten slotted nut to 200 N·m (145 ft.-lbs.) torque. Tighten clamp bolt to 40 N·m (30 ft.-lbs.) torque.

DRIVE SHAFT

7. To remove the drive shaft, first unbolt front and rear antiwrapping guards and slide them toward center of drive shaft. Unbolt and separate yoke flange (1—Fig. 6) from drive hub on front differential pinion shaft. Pull drive shaft with antiwrapping guards forward off the clutch shaft and remove from tractor.

Check "U" joints for excessive wear and renew as necessary. "U" joint cross (3), seals, bearing needles and bearing sleeves are available as an assembly with new snap rings (2). Removal and installation of "U" joints is conventional.

Reinstall drive shaft assembly by reversing removal procedure. Tighten yoke flange bolts to a torque of 75 N·m (55 ft.-lbs.).

FRONT DRIVE AXLE AND SUPPORT

8. REMOVE AND REINSTALL. To remove the front drive axle and support assembly, remove drive shaft as outlined in paragraph 7. Disconnect power steering lines at steering cylinder. Cap or plug openings immediately to prevent dirt from entering system. Support tractor behind axle carrier and remove front wheels. Support front axle assembly with a suitable jack or special safety stand to prevent tipping while permitting the axle to be moved safely. Refer to Fig. 7 and remove cotter pin, slotted nut (1) and washer (2). Remove special pivot bolt (6), washer (2) and shim (3). Move front axle and support (5) rearward until clear of pivot tube (4) and rear pivot pin. Lower axle assembly and carefully roll away from under tractor.

Axle pivot support (5) can be unbolted and removed from front drive axle if necessary. If either the support or the axle housing is renewed, it will be necessary to ream the holes in the new part to 21 mm (0.826 inch) and in alignment with holes in the mating part. The four cap screws (two 20 × 75 mm and two 20 × 160

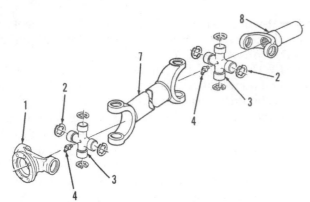

Fig. 6—Exploded view of front drive shaft and "U" joints used on models with front-wheel drive.

1. Yoke flange	4. Grease fitting
2. Snap rings	7. Drive shaft
3. "U" joint cross	8. Slip yoke

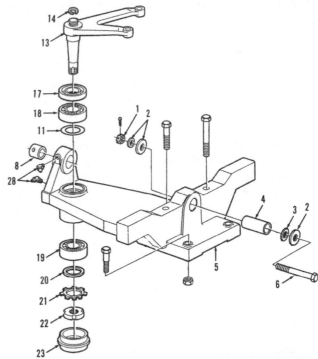

Fig. 7—Pivot support used on 2150 Models equipped with front-wheel drive. Support is bolted to top of front drive axle housing.

1.	Slotted nut	14.	Snap ring
2.	Washer	17.	Sealing disk
3.	Shim	18.	Upper bearing
4.	Pivot tube	19.	Lower bearing
5.	Support	20.	Washer
6.	Pivot bolt	21.	Lock washer
8.	Bushing	22.	Nut
11.	Shim	23.	Cap
13.	Steering bellcrank	28.	Grease fittings

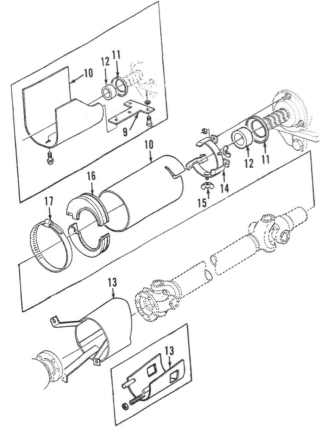

Fig. 6A—The drive shaft to the front drive axle is protected by the anti-wrapping guard. Different styles have been used.

9.	Bracket		
10.	Rear guard	14.	Bracket
11.	Washer	15.	Wing nuts
12.	Bushing	16.	Spacer
13.	Front guard	17.	Clamp

mm) and two nuts (on special 20 × 90 mm alignment screws) should be tightened to 480 N·m (355 ft.-lbs.) torque when assembling. Bellcrank (13) is supported in tapered roller bearings, which should be lubricated by packing with EP multi-purpose grease. The tapered roller bearings should be correctly located vertically by adding the required number of shims (11) between the bearing cup of upper bearing and bore in axle center member. Correct selection of shims is determined by measuring the depth of bearing bore, then subtracting 18.4 mm (0.724 inch). The result is thickness of shims (11) required. Shim thickness can be rounded to the nearest tenth (0.1) mm when assembling. Tighten the special nut (22) to 55 N·m (40 ft.-lbs.) torque, then lock position with tab of washer (21). Check bushing (8), front pivot tube (4), rear pivot pin (7—Fig. 2) and bushing (24) for excessive wear and renew if necessary. Be sure to align passage in bushing with appropriate grease fitting if new bushing is installed. Crossed grooves in bushing (8—Fig. 7) should be down and slot in bushing (24—Fig. 2) should be up. Reverse removal procedure when assembling. Tighten slotted nut (1—Fig. 7) to 300 N·m (220 ft.-lbs.) torque. End clearance of axle center member should be 0-0.5 mm (0-0.020 inch) and is adjusted by adding or removing shims (3). If cotter pin cannot be installed through slotted nut (1) after tightening to specified torque, tighten enough to align hole with next slot, then install cotter pin.

WHEEL HUB WITH PLANETARY

9. REMOVE AND REINSTALL. To remove front wheel hub and planetary as an assembly with axle, block up under axle housing and remove front wheel. Attach a suitable hoist to support hub assembly. Unbolt hub carrier (6—Fig. 8) from steering knuckle housing (20—Fig. 11). Withdraw hub assembly with axle shaft (38 through 41—Fig. 8).

Reinstall hub and axle assembly being careful not to damage seal ring and oil seal (14 and 13—Fig. 11). Tighten cap screws securing hub carrier to steering knuckle housing to a torque of 190 N·m (140 ft.-lbs.). Reinstall wheel and tighten lug nuts to 300 N·m (220 ft.-lbs.) torque.

10. OVERHAUL. To disassemble the front wheel hub and planetary, block up under axle housing and remove front wheel. Remove drain plug (30—Fig. 8) and drain lubricant from hub. Remove cap screws from outer perimeter of planetary carrier (24). Using two special jackscrews #KML10013, remove planetary carrier from hub (11). Remove snap rings (21) and drive out planetary shafts (25). Carefully remove planetary gears (35) with thrust washers (33 and 37), spacers (34 and 36) and bearing needles (32). Remove

snap ring (20) and pull sun gear (19) from end of axle shaft (1). Straighten lock plate (17), then using special ring nut socket #KML10011, remove nut (18), lock plate (17) and nut (16). Pull ring gear assembly (13, 14 and 15) from hub. Remove hub (11) with bearing assembly (12) from hub carrier (6). Remove bearing cups from hub if necessary. Remove bearing cone (9), retainer (8) and oil seal (7) from carrier (6). Unbolt and separate hub carrier from steering knuckle housing. Remove retainer (2), seal ring (3),

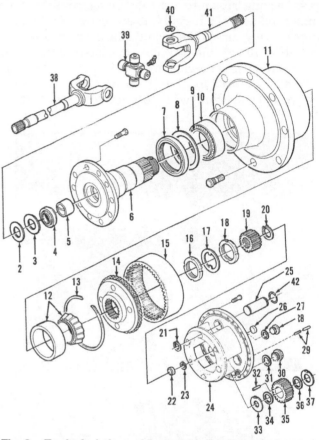

Fig. 8—Exploded view of front wheel hub and planetary used on front-wheel drive equipped 2150 Models.

2.	Retainer		
3.	Seal ring	23.	Shim
4.	Oil seal	24.	Planetary carrier
5.	Bushing	25.	Shaft
6.	Hub carrier	26.	Plug
7.	Oil seal	27.	Seal ring
8.	Retainer	28.	Level plug
9.	Bearing cone	29.	Roll pins
10.	Bearing cup	30.	Drain plug
11.	Wheel hub	31.	Seal ring
12.	Tapered roller bearing	32.	Bearing needle
13.	Snap ring	33.	Thrust washer
14.	Ring gear carrier	34.	Spacer
15.	Ring gear	35.	Planetary gear
16.	Adjusting nut	36.	Spacer
17.	Lock plate	37.	Thrust washer
18.	Lock nut	38.	Inner axle shaft
19.	Sun gear	39.	"U" joint cross
20.	Snap ring	40.	Snap rings
21.	Snap ring	41.	Outer shaft
22.	End stop	42.	Snap ring

oil seal (4) and bushing (5) if damaged or worn. Axle shaft (38) can be pulled from axle housing at this time.

Clean and inspect all parts and renew any showing excessive wear or other damage.

When reassembling, carefully slide axle shaft (38) into axle housing until engaged in differential side gear. If removed, install bushing (5), oil seal (4), seal ring (3) and retainer (2) in hub carrier (6). Carefully install hub carrier over axle shaft and tighten cap screws securing hub carrier to steering knuckle housing to a torque of 190 N·m (140 ft.-lbs.). Install oil seal (7), retainer (8) and bearing cone (9). If removed, press bearing cups into hub (11). Being careful not to damage oil seal (7), install hub on carrier (6). Install outer bearing cone, then install ring gear and carrier assembly (13, 14 and 15). Install adjusting nut (16) and tighten using special ring nut socket until a pull of 30-49 N (6.5-11 lbs.) on a spring scale with cord wrapped around hub, will keep hub rotating. Install

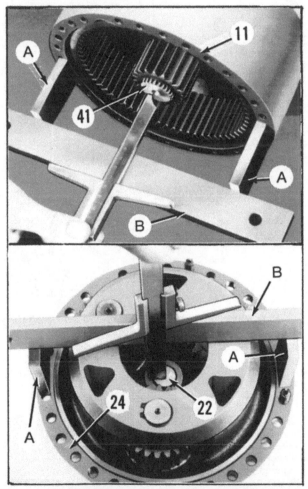

Fig. 9—Measure the distance between end of outer axle (41) and face of hub (11), then measure distance between flange of planet carrier (24) and end stop (22). Difference is end play of axle shaft. The thickness of gage blocks (A) and straight edge is not important as long as measurement is possible and the same gage blocks are used for both measurements. Change end play by varying thickness of shims (23—Fig. 8).

lock plate (17) and tighten locknut (18). Secure nuts with lock plate. Install sun gear (19) and snap ring (20). Reinstall planetary gears (35), bearing needles (32), spacers (34 and 36) and thrust washers (33 and 37) in planetary carrier (24). Insert shafts (25) and secure with snap rings (21). Before installing planetary carrier assembly, determine and adjust axial play of axle shaft as follows: Turn steering knuckle housing up against stop and make sure that axle shaft turns freely. In this position, measure and note distance between outer face of hub (11—Fig. 9) and outer end face of axle shaft (41), using a depth gage, gage blocks and cross bar. Using the same depth gage and gage blocks, measure distance between flange of planetary carrier (24) and end stop (22). Check the difference between the two measurements against the specified end clearance of 0.3-0.5 mm (0.012-0.020 inch). If necessary, add or remove shims (23—Fig. 8) behind end stop (22) to correct the end clearance.

Install planetary carrier and tighten cap screws to a torque of 35 N·m (25 ft.-lbs.). Fill hub with EP SAE 90 or equivalent oil to level of plug opening marked "OLSTAND" with plug hole and marking horizontal. Capacity of each front hub is 0.75 L (0.79 quart). Reinstall wheel assembly and tighten lug nuts to 300 N·m (220 ft.-lbs.) torque. If axle was removed, be sure to check and fill axle center housing to level of plug in front center of housing. Recommended lubricant is EP SAE 90 or equivalent.

STEERING KNUCKLE HOUSING

11. REMOVE AND REINSTALL. To remove either steering knuckle housing, first remove the wheel hub with planetary as outlined in paragraph 9. Disconnect tie rod and steering cylinder from steering knuckle. Unbolt and remove steering arm (17—Fig. 11) from right side. Remove attaching screws and lock plates (16 and 23), then remove grease fitting and extension, if used. Heat area of knuckle housing around king pins (19 and 21) to release Loctite No. 277, then unscrew king pins.

Remove steering knuckle housing (20), then remove "O" rings (7), dust seals (8) and bearing cones (9) from axle housing fork. Drive bearing cups (10) and grease cups (11) out of axle housing fork if renewal is required.

Clean and inspect all parts and renew any showing excessive wear or other damage. Inspect bushing, oil seal, seal ring and retainer (12, 13, 14 and 15) for wear or damage and renew as necessary.

When reassembling, install grease cups (11) and bearing cups (10). Pack bearing cones (9) with EP multi-purpose grease and install bearing cones and dust seals (8) in axle housing fork. Be sure threads are clean, then coat threads for king pins (19 and 21) with Loctite 277. Place steering knuckle housing (20) in position on axle housing and install "O" rings (7) and king pins (19 and 21). Be careful when installing king pins to be sure that dust cap (8) and lower bearing cone (9) stay in position until king pins are fully installed. Tighten king pins evenly on top and bottom, then adjust bearing load until rolling torque to rotate knuckle (20) is 11-15 N.m (8-11 ft.-lbs.). Install lock plates (16 and 23) when adjustment is complete. Tighten screws securing steering arm to 230 N.m (170 ft.-lbs.) torque.

Reconnect steering cylinder and tie rod, then reinstall wheel hub assembly as outlined in paragraph 9.

DIFFERENTIAL UNIT

12. R&R AND OVERHAUL. To remove the differential assembly, first remove drain plug and drain lubricant from axle housing. Block up under both ends of axle housing and remove both front wheel hub assemblies as outlined in paragraph 9. Remove drive shaft as outlined in paragraph 7. Unbolt and lift complete differential unit from front axle housing.

Remove roll pins (17—Fig. 12), then remove both adjusting rings (12). Remove cups for both bearings (11), then use a suitable puller to remove bearing cones from differential case. Remove the ten special screws (28), then carefully lift the differential from the carrier (10). Notice that the screws (28) hold ring gear (18), cover (19) and case (27) together.

Separate differential case (27), cover (19) and ring gear (18). Lift out the side gear (23) and clutch packs (20, 21 and 22) from the open side and keep these parts separate from similar parts for opposite side. Remove snap ring (1), withdraw pinion shaft (26), pinions (24) and thrust washers (25). The second side gear (23) and clutch pack (20, 21 and 22) can then be removed. Although parts (20, 21, 22 and 23) are alike for both sides, parts should be kept together if they are to be reinstalled.

Heat pinion nut (2) to release Loctite 277, then hold hub (4) and remove nut (2). Pull hub (4) from pinion shaft splines, then press pinion out of carrier (10) and bearing cone (7). Shims (13) are used to set position of pinion (16) and should be correct if bearing (14 and 15) and pinion are reinstalled in same housing. Crush spacer (9) is used to set rolling torque of pinion shaft (16) in bearings (7, 8, 14 and 15) and should be new when assembling pinion in housing.

Clean and inspect all parts and renew any that show excessive wear or other damage. Always renew ring gear (18) and pinion (16) as a matched set.

To assemble pinion (16) and associated parts in carrier housing and check pinion mesh position, proceed as follows:

Install bearing cup (14) in bore of housing, making sure that cup is fully seated, then position bearing (15) squarely in cup. Measure distance from the bearing cone (15) and the center line of carrier bearing

bores in housing using special John Deere tools (JDG-74-1 and JDG-74-2) or equivalent as shown in Fig. 13. Compare this result with the measurement etched on face of drive pinion. Compensate difference of measurements by adding or removing shims (13—Fig. 12). Remove special tools, then install bearing cup (8). Heat bearing cones (7 and 15) before pressing into position. Press bearing cone (15) on drive pinion (16) until bottomed against shoulder. Install drive pinion in carrier, then install crush spacer (9) and press bearing cone (7) on drive pinion. **DO NOT** install oil seal (6) at this time. Install drive flange (4), washer (3) and nut (2). Tighten nut and **note the tightening torque required** on nut to obtain a rolling torque on drive pinion of 1-2 N.m (9-18 in.-lbs.). Remove nut, washer and drive flange, then install oil seal (6) and dust shield (5). Install drive flange and washer, coat threads of nut with Loctite 277, then install and tighten nut to previously noted torque.

Assemble differential as follows: If previously used parts are to be reinstalled, they should be located in their original locations or abnormal wear may result. Thickness of clutch packs (20, 21 and 22) should be exactly the same thickness on both sides of pinion gears (24). When measuring thickness of clutch packs and measuring depth of clutch pack in differential case, parts should be **dry.**

Assemble alternating five external lugged plates (20) and four internal splined discs (21), then install one compensating disc (22) and side gear (23) in differential case. Position thrust washers (25) and pinion gears (24) in case, then insert shaft through

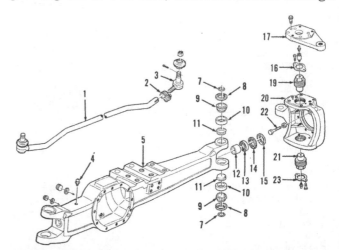

Fig. 11—Exploded view of steering knuckle housing, front drive axle housing and tie rod used on 2150 Models.

1. Tie rod	
2. Clamp	13. Oil seal
3. Tie rod end	14. Seal ring
4. Vent	15. Retainer
5. Axle housing	16. Lock plate
7. "O" ring	17. Steering arm
8. Dust seal	19. King pin
9. Bearing cone	20. Steering knuckle housing
10. Bearing cup	21. King pin
11. Cup	22. Stop adjusting screw
12. Bushing	23. Lock plate

case, thrust washers and pinion gears. Position second side gear (23) in case with teeth meshed with pinion gears, then assemble one compensating disc (22) and the second clutch pack. Be sure that compensating disc (22) on each side gear is the same thickness and that the thickness of the clutch packs (21 and 22) are the same when assembled dry. Measure depth from face of differential case to last plate (20) as shown at (A—Fig. 14). Measure distance from plate contacting surface of cover to the mating flange of cover as shown at (B). Subtract the measurements to determine axial clearance of clutch pack. Recommended clearance should be 0.10-0.20 mm (0.004-0.008 inch) and is changed by varying the thickness of the compensating discs (22—Fig. 12). Discs (22) are available in thicknesses of 2.8, 2.9 and 3.0 mm (0.110, 0.114 and 0.118 inch). The measurement is of only one side, but changes should be made to both sides and the combined thickness of clutch pack parts (20, 21 and 22) must be exactly the same. Reassemble and measure to be sure that proper clearance is maintained. After correct clutch pack clearance is determined, disassemble, coat all parts with SAE 90

Hypoid gear oil and reassemble. Install pinion shaft snap ring (26) and cover (19). Screws (28) should be tightened evenly to 100 N•m (74 ft.-lbs.) torque.

Fig. 13—Measure distance from bearing (15) to center line of carrier bearing bores in housing (10) using special tools as shown. Correct measurement is etched on end of pinion.

1. Snap ring
2. Nut
3. Washer
4. Drive hub
5. Dust shield
6. Oil seal
7. Bearing cone
8. Bearing cup
9. Crush spacer
10. Differential carrier
11. Tapered roller bearing cone & cup
12. Adjusting rings
13. Shim
14. Bearing cup
15. Bearing cone
16. Drive pinion
17. Roll pins
18. Ring gear
19. Differential case cover
20. External lugged plates
21. Internal splined discs
22. Compensating discs
23. Bevel side gears
24. Bevel pinions
25. Thrust washers
26. Pinion shaft
27. Differential case
28. Special screws

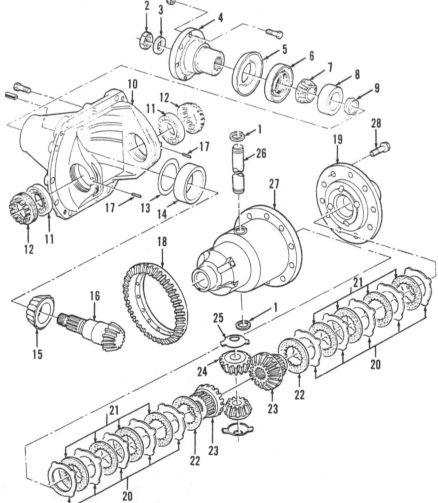

Fig. 12—Exploded view of front drive axle differential unit used on 2150 Models.

14

Position assembled differential case in carrier housing (10) after installing and adjusting pinion. If bearing cones (11) are installed, it may be necessary to remove some screws (28). Heat and install cones of differential carrier bearings (11). Install bearing cups in differential carrier (10). Spray threads of adjusting rings (12) with Molycote. Install adjusting rings and tighten against bearings (11) until differential end play is zero. Turn adjusting rings as necessary to obtain backlash of 0.13-0.20 mm (0.005-0.008 inch) between ring gear and drive pinion.

NOTE: When adjusting backlash, both adjusting rings must be turned the same amount so that end play remains at zero.

To obtain specified rolling torque of 1-2 N·m (9-18 in.-lbs.), turn adjusting ring opposite the ring gear two more notches. Recheck ring gear backlash, which should be 0.13-0.20 mm (0.005-0.008 inch). Install roll pins (17) to secure adjusting rings.

Using a dial indicator, check ring gear run-out. Maximum allowable run-out is 0.08 mm (0.003 inch). Check tooth contact pattern as follows: Coat several ring gear teeth with bluing. Rotate drive pinion forward and backward and inspect tooth contact pattern. See Fig. 15. If ideal contact pattern is not made, complete measuring process must be repeated.

Reassemble by reversing the disassembly procedure. Tighten differential carrier retaining cap

screws to a torque of 70 N·m (50 ft.-lbs.). Fill front axle housing to level plug opening with EP SAE 90 transmission oil or equivalent. Capacity is 5.25 L (5.6 quarts).

FRONT-WHEEL DRIVE CLUTCH

13. R&R AND OVERHAUL. To remove the front-wheel drive disc clutch assembly, first drain transmission oil. Disconnect wires from solenoid. Unbolt and remove oil pan and solenoid assembly from clutch housing. Remove the drive shaft as outlined in paragraph 7. Remove washer and plastic sleeve from front end of shaft (24—Fig. 26). Unbolt and remove bearing quill (30) with seal washer (29), oil seal (28), "O" ring (27) and shim (26). Install a cap screw and washer in front end of clutch shaft. Attach a slide hammer puller and pull clutch shaft forward until front bearing cup is free of clutch housing. Support disc clutch assembly with one hand and pull clutch shaft out the front.

NOTE: When clutch shaft is removed, rear bearing cone and spacer are free to drop out of clutch housing.

Lower disc clutch assembly from clutch housing and catch bearing cone and spacer.

To disassemble the clutch, refer to Fig. 26, then unbolt and remove covers (3 and 20). Pull clutch hub (5) from clutch assembly. Tilt clutch assembly and remove thrust washer (16). Use a JDT-24A spring compressor or place assembly in a press and apply pressure to Belleville springs, then remove snap ring (6). Release pressure and remove Belleville springs (7), rear pressure ring (8), internal and external splined discs (9 and 10) and center pressure ring (11). Remove snap ring (12), release ring (13), front pressure ring (8), piston (14) with piston ring (15) and seal

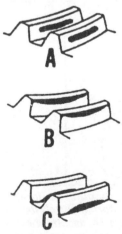

Fig. 15—Views of typical tooth contact patterns on ring gear.

A. Ideal tooth contact pattern
B. Drive pinion must be moved toward ring gear
C. Drive pinion must be moved away from ring gear

Fig. 14—Difference between measurements (A and B) is used for setting the operating clearance of the self locking clutch in the front differential of 2150 Models.

Illustrations for Fig. 14 and Fig. 15 reproduced by permission of Deere & Company. Copyright Deere & Company.

ring (17) from clutch drum (18). Remove front bearing cone (25) and "O" rings (19) from clutch shaft. Remove rear bearing cup (1) from clutch housing, if necessary.

Clean and inspect all parts and renew any showing excessive wear or other damage. Renew all "O" rings and seal rings when reassembling. Press new bushing (4) in clutch hub (5), if necessary.

To reassemble clutch, insert seal ring (17) in groove of clutch drum (18). Install piston ring (15) on piston (14), then install piston in clutch drum. Install front pressure ring (8), release ring (13) and secure with snap ring (12). Install center pressure ring (11) with chamfered side facing snap ring. Starting with internal splined disc (9), alternately install **eight** internal splined and **seven** external splined discs (10). Install rear pressure ring (8), then four Belleville springs (7) as shown at (B—Fig. 27). Make certain that internal splines of clutch discs are correctly aligned. Using spring compressor JDT-24A or a press, compress Belleville springs and install snap ring (6—Fig. 26). Install thrust washer (16) with chamfered side facing clutch drum. Carefully insert clutch hub and gear into clutch discs. Install covers (3 and 20) and secure with three cap screws (21).

Before installing clutch assembly, check and adjust, if necessary, clutch shaft axial free play. Install rear bearing cup (1) in clutch housing. Install clutch shaft (24) with rear bearing cone (1), spacer (2) and front bearing (25). Install bearing quill (30) without shims (26) and secure with three cap screws (31). With a dial indicator against front end of clutch shaft, move shaft forward and rearward by hand and check axial end play. Deduct the specified axial free play of 0.02-0.07 mm (0.001-0.003 inch) from the measured end play to determine correct thickness shim pack to be installed. Remove bearing quill, shaft and bearings. Remove rear bearing cone (1) and spacer (2) from shaft. Use light grease to stick bearing cone in bearing cup (1) that is already in clutch housing. Install new seal ring

(23) on sleeve (22) and new "O" rings (19) in grooves of shaft at each side of hydraulic oil port. Hold clutch assembly into clutch housing and insert clutch shaft until spacer (2) can be installed. Install spacer with chamfered side facing bearing (1). Push shaft fully to the rear and install front bearing cup (25). Install the determined shim pack (26) and new "O" ring (27). Install bearing quill (30), with seal washer (29) and oil seal (28) and secure with cap screws (31).

Reinstall remaining parts by reversing removal procedures. Fill transmission to level mark on dipstick. Install drive shaft as outlined in paragraph 7.

CLUTCH OIL PRESSURE TEST

14. To check front-wheel drive clutch oil pressure, operate tractor until transmission oil temperature is 60° C (140° F). Remove test plug (5—Fig. 19) and install a 2000 kPa (300 psi) test gage.

> **CAUTION: If the plug has 9/16 or 5/8 inch head, the threads should be English; if plug removal requires 3/4 inch or metric size wrench, port probably has metric threads. If there is any question, check thread size before attaching test gage.**

With engine operating at 1000 rpm and front-wheel drive control switch in "Disengaged" position, operating pressure should be approximately 1050 kPa (150 psi). With control switch in "Engaged" position, pressure gage should read zero. If not, repair solenoid as outlined in the following paragraph.

To check for internal leakage, such as caused by leaking clutch piston seals, install two (matched) 2000 kPa (300 psi) gages as described in Fig. 16. With transmission in "Park," dash "MFWD" switch OFF and engine operating at 1000 rpm, pressure at the two test points should be within 170 kPa (25 psi) of the same pressure. If special test cover No. 0745 is used instead of "T" fitting shown in Fig. 16, maximum allowable difference can be 185 kPa (27 psi).

SOLENOID

15. When front-wheel drive control switch is in "Disengaged" position, solenoid is energized causing solenoid spool to direct pressurized oil to clutch piston. Clutch piston compresses the Belleville springs disengaging the clutch. When control switch is in "Engaged" position, there is no current to the solenoid and a spring forces spool to the off position. In this position, there is no oil pressure at clutch piston and the Belleville springs engage the clutch.

To remove the solenoid, remove drain plug (1—Fig. 19) and drain oil. Unbolt and remove solenoid guard (8). Disconnect wires from solenoid. Remove "E" ring (10 or 22) and pull solenoid coil (11 or 25) from core

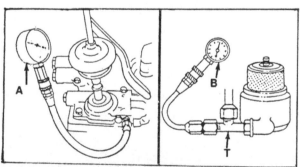

Fig. 16—Two gages should be installed at locations shown to check for internal clutch leakage on 2150 Models with MFWD. Gage (A) is attached to port shown in clutch control valve housing and gage (B) attached at filter. The tee (T) or the special ported filter cover must be installed to attach gage (B). Since difference in pressure at the two ports is measured, it is important that the gages are matched to indicate exactly the same pressure readings.

(14 or 31). Remove Allen screws (12 or 27) and retaining plate (13 or 28). Valve parts can be pulled from bore in pan if cleaning, repair or renewal is required. Early and late parts are different, but operation and service procedures are similar.

Clean and inspect all parts for wear or damage and renew as necessary. Spool and sleeve are available only as a matched set. Use new back-up washer and "O" rings when reassembling and reinstall by reversing the disassembly and removal procedures.

FRONT-WHEEL DRIVE SYSTEM (2155, 2350, 2355, 2355N, 2550 AND 2555 MODELS SO EQUIPPED)

A mechanical front-wheel drive is available on some 2155, 2350, 2355, 2355N, 2550 and 2555 tractors. There are some differences between the Front-Wheel Drive Systems used on these models that will be referred to in the servicing and testing instructions that follow. Please note the following applications:

Tractor Model	Drive System Model
2155	APL315
2350	APL325
2355	APL325
2355N	APL315N
2550	APL325
2555	APL325

The front-wheel drive is controlled by an electric solenoid/hydraulic valve that operates a multiple disc clutch. The clutch is engaged mechanically by preloaded Belleville springs and is disengaged hydraulically. The clutch is located in the bottom front of the synchronized transmission case. A drive shaft with two "U" joints connects the clutch unit to front axle. Front axle is equipped with a self-locking differential and a planetary drive at each of the front wheel hubs. The gears that drive the clutch are integral with the synchronized transmission.

CAUTION: When servicing front-wheel drive equipped tractor with rear wheels supported off ground, engine running and transmission in gear, always support front wheels off ground too. Loss of electrical power or transmission hydraulic system pressure will engage front driving wheels. Rear wheels will be pulled off support if front wheels are not raised.

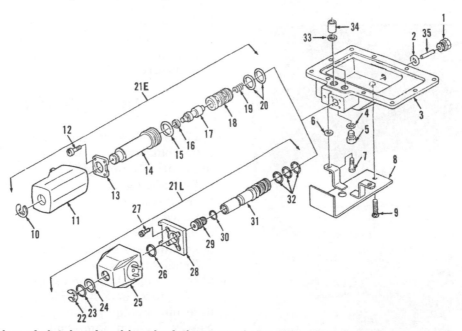

Fig. 19—Exploded view of clutch solenoid and relative parts. Later solenoid/valve assembly is shown at 21L and early type is at 21E.

1. Drain plug		
2. Seal ring	9. Cap screw	16. Back-up washer
3. Oil pan	10. "E" ring	17. Spool
4. "O" ring	11. Solenoid coil	18. Sleeve
5. Test plug	12. Allen screw (4)	19. Spring
6. Washer	13. Retaining plate	20. "O" rings
7. Cap screw (2)	14. Solenoid core	21E. Solenoid/valve unit
8. Solenoid guard	15. "O" ring	21L. Solenoid/valve unit

22. "E" ring	29. Plug
23. "O" ring	30. "O" ring
24. "O" ring	31. Core sleeve
25. Solenoid coil	32. "O" rings
26. "O" ring	33. Seal rings (2)
27. Allen screw (4)	34. Tube
28. Retaining plate	35. Pin (8 × 36 mm)

TROUBLE-SHOOTING

16. Some problems that may occur with front-wheel drive and their possible causes are as follows:

1. Front-wheel drive not engaging. Could be caused by:
   a. Dash electrical switch defective. Check to be sure that electrical circuit is open when key switch and dash "MFWD" switch are both ON. (The engine does not need to be running when making this check.) Outside of control solenoid should not have strong magnetic field as checked by placing a screwdriver near solenoid.
   b. Solenoid/hydraulic valve spool sticking open. The valve may stick if the valve attaching screws are improperly or unevenly tightened.
   c. Disc clutch pack worn. To check for slippage with engine stopped, set parking brake, slide the drive shaft shield forward and remove the rear guard mounting bracket. Raise the left front wheel off ground using a floor jack, then use a pipe wrench or other suitable tool attached to the clutch output shaft to measure torque required to cause clutch to slip. Clutch should not slip at less than 880 N·m (650 ft.-lbs.) torque for all models with APL315, APL315N and APL325 Front-Wheel Drive Systems. Torque can be

measured using a pipe wrench and suitable handle extending pipe by dividing your total body weight into the desired torque and the result will be the distance from shaft center line to apply your weight. An example would be 650 ft.-lbs. (the desired torque) ÷ 180 lbs. (your body weight) = 3.61 ft. (distance from shaft center line). To convert the distance to inches, multiply $3.61 \times 12 = 43.32$ inches.
   d. Mechanical failure in front-wheel drive.

2. Front-wheel drive not disengaging. Can be checked by shifting transmission to "Neutral," then jacking up left front wheel. With key switch ON, dash "MFWD" switch OFF, engine operating at 1000 rpm and oil temperature warmed to at least 38° C (100° F), it should be possible to rotate the left front wheel by hand. Failure to disengage could be caused by:
   a. Electrical fault (switch, fuse or wiring). Check for magnetic field at solenoid valve using a screwdriver or similar object near the valve with key switch ON and dash "MFWD" switch OFF.
   b. Defective solenoid/hydraulic valve (21E or 21L— Fig. 19). The valve may stick if the valve attaching screws (12 or 27) are improperly or unevenly tightened.
   c. Hydraulic pressure too low or leaking clutch piston sealing rings. Pressure can be checked as outlined in paragraph 24.

3. Excessive front tire wear. Could be caused by:
   a. Toe-in incorrect.
   b. Front-rear tire combination not as specified.
   c. Incorrect tire inflation.

TIE RODS AND TOE-IN

17. Tie rod ends may be one of several different types, but none are adjustable for wear and faulty units must be renewed.

To check toe-in, first turn steering wheel so that front wheels are in straight ahead position. Measure distance at front and rear of front wheels from rim flange to rim flange at hub height. Toe-in should be 0-3.0 mm (0-⅛ inch).

To adjust toe-in on models with tie rod end shown at top of Fig. 20, loosen clamp bolt (A) on tie rod and turn adjuster end (B), which is nearest to piston rod, in or out as required. Adjust tie rods on both sides equally. Tighten clamp bolt (A) to a torque of 50 N·m (35 ft.-lbs.) when adjustment is correct.

To adjust toe-in on models with tie rod end shown at bottom of Fig. 20, loosen clamp bolt (D), remove cotter pin and slotted nut (C). Disconnect tie rod end from steering arm, then turn tie rod end in or out of tie rod tube as required to correctly adjust toe-in. Reattach tie rod end to steering arm and recheck toe-in after making changes. When adjustment is

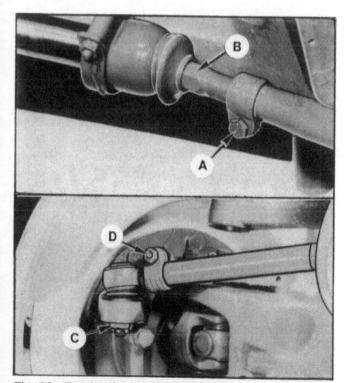

Fig. 20—Toe-in of models equipped with steering cylinder integral with axle housing is adjusted by turning tie rod end (B) with 22 mm wrench. Toe-in of models with tie rod ends as shown in lower view should be disconnected from steering arm and threaded in or out of tie rod tube.

correct, tighten slotted nut (C) to 180 N·m (130 ft.-lbs.) torque and install cotter pin. Tighten nut for clamping bolt (D) to 40 N·m (30 ft.-lbs.) torque.

DRIVE SHAFT

18. To remove the drive shaft, first unbolt front and rear antiwrapping guards (Fig. 6A) and slide them toward center of drive shaft. Unbolt and separate yoke flange (1—Fig. 6) from drive hub on differential drive pinion shaft. Pull drive shaft with antiwrapping guards forward off the clutch shaft and remove from tractor.

Check "U" joints for excessive wear and renew as necessary. "U" joint cross (3), seals, bearing needles and bearing sleeves are available as an assembly with new snap rings (2). Removal and installation of "U" joints are conventional.

Reinstall drive shaft assembly by reversing removal procedure. Tighten yoke flange bolts to a torque of 75 N·m (55 ft.-lbs.).

FRONT DRIVE AXLE

19. REMOVE AND REINSTALL. To remove the front drive axle assembly, remove drive shaft as outlined in paragraph 18. Disconnect power steering lines at steering cylinder. Cap or plug openings immediately to prevent dirt from entering system. Support tractor behind axle carrier and remove front wheels. Support front axle assembly with a suitable jack or special safety stand to prevent tipping while permitting the axle to be moved safely. Refer to Fig. 22 and remove cotter pin (6), slotted nut (7) and washers (8 and 9). Remove pivot bolt (13), washer (12) and shim (11). Move axle assembly rearward until clear of pivot tube (10) and rear pivot pin. Lower axle assembly and carefully roll away from under tractor.

Some models are equipped with a bellcrank (35—Fig. 20A) that transmits longitudinal movement of steering drag link to lateral movement at steering arms. Cups for tapered bearings (33 and 36) should be pressed into bellcrank until seated against snap ring (34). Bearing cones should be packed with EP multi-purpose grease. Pin (39) is pressed into bore of housing (3). Position washer (38) over pin, coat space between sealing edges of cover disc (37) with grease and position disc (37) over pin with chamfered edge up. Install lower bearing cone over pin, position bellcrank and upper bearing cone over bellcrank, then install lock washer (32) and nut (31). Tighten nut (31) to 60 N·m (45 ft.-lbs.) torque, lock with washer (32), fill dust cap (30) with grease, then install dust cap.

On all models, inspect bushing (4—Fig. 22), pivot tube (10), bushing (24—Fig. 2) and pin (7) for excessive wear and renew if necessary. Be sure to align passage in bushing with hole from grease fitting if new bushing is installed. Crossed grooves of rear

bushing (4—Fig. 20A or Fig. 22) should be down and crossed grooves in front bushing (24—Fig. 2) should be up.

Reinstall front drive axle assembly by reversing the removal procedure. Tighten slotted nut (7—Fig. 20A or Fig. 22) to a torque of 300 N·m (220 ft.-lbs.), then check end clearance of drive axle assembly. Add or remove shims (11) as required to obtain end clearance of 0-0.5 mm (0-0.020 inch).

WHEEL HUB AND PLANETARY

20. R&R AND OVERHAUL. To remove either front wheel hub and planetary, support front axle housing and remove front wheel. Remove drain plug (42—Fig. 21) and drain oil from hub assembly. Remove the two Allen screws (41) and lift off planetary carrier (40). Remove thrust washers (39), then remove three snap rings (29). Using a suitable puller, remove planet gears (32) with bearings (31). Remove angular snap rings (30 or 33) and separate bearings (31) from planet gears. Remove snap ring (36) and

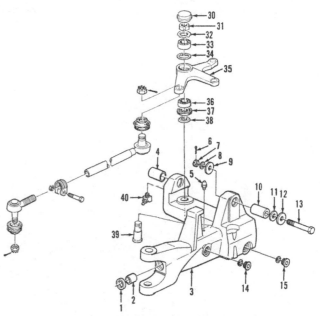

Fig. 20A—Exploded view of steering bellcrank used on some Four-Wheel-Drive models. Bellcrank pin (39) is 94.5 mm (3.72 inches) long for earlier models, 91.5 mm (3.60 inches) long for later models. Pin is located in bore of the left front drive axle housing (3).

1. Oil seal	14. Level plug
2. Bushing	15. Drain plug
3. Axle housing R. H.	30. Dust cap
4. Bushing	31. Nut
5. Vent	32. Locking washer
6. Cotter pin	33. Upper bearing
7. Slotted nut	34. Snap ring
8. Washer	35. Bellcrank
9. Washer	36. Lower bearing
10. Pivot tube	37. Cover disc
11. Shim	38. Washer
12. Washer	39. Pin
13. Cap screw	40. Grease fitting

thrust washer (34) from pinion shaft and sun gear (35).

Before removing the ring gear (28), mark relative position of ring gear (28) and steering knuckle housing (17) to facilitate installation in same location. Unscrew the eight self-locking cap screws (38). Install three or four M10 × 80 mm cap screws in holes around sun gear and using a disc plate (OTC 27536 or equivalent) against cap screw heads, attach a jawed puller to ring gear (28) with center pressing against disc plate that is against the M10 × 80 mm screws. Heat area of ring gear around bushings (37) to 300° C (570° F) and remove ring gear. Bushings (37) are installed with Loctite and heat must be used to release the

Loctite. Use a slide hammer puller and remove bushings (37).

To remove the hub (25) and bearing cone (26) from steering knuckle (17), reinstall the three or four M10 × 80 mm cap screws around sun gear and use a disc plate (OTC 27536 or equivalent) against heads of screws. Attach a large jawed puller 310 mm (12.2 inches) across hub with center pressing against disc plate that is against heads of screws and pull hub (25) from steering knuckle housing (17). Remove "O" ring (27) and outer bearing cone (26) from hub, then remove inner bearing cone (23), oil seal (22) and scraper ring (21) from knuckle housing (17). If necessary,

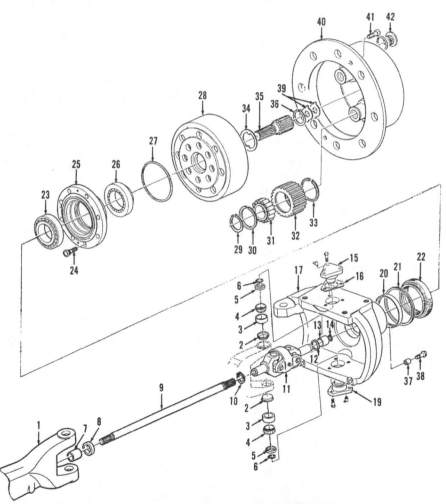

Fig. 21—*Exploded view of front wheel hub, planetary and steering knuckle housing used on some models equipped with front-wheel drive.*

1. Axle housing
2. Cup
3. Bearing cup
4. Bearing cone
5. Seal cap
6. "O" ring
7. Bushing
8. Oil seal
9. Axle shaft
10. Snap ring
11. Double "U" joint
12. Oil seal
13. Bushing
14. Snap ring
15. King pin
16. Shim
17. Steering knuckle housing
19. King pin
20. Wear ring
21. Scraper ring
22. Oil seal
23. Taper roller bearing
24. Lug bolt
25. Hub
26. Taper roller bearing
27. "O" ring
28. Ring gear
29. Snap ring
30. Angular snap ring
31. Roller bearing
32. Planetary gear
33. Angular snap ring
34. Thrust washer
35. Pinion shaft & sun gear
36. Snap ring
37. Bushing
38. Cap screw
39. Shim washers
40. Planet carrier
41. Cap screw
42. Drain plug

remove wear ring (20) from knuckle housing and bearing cups from hub.

Clean and inspect all parts for excessive wear or other damage and renew as necessary.

When reassembling, drive bearing cups in hub, then install inner bearing cone, oil seal (22) and scraper ring (21) in hub (25). If removed, install new wear ring (20) on knuckle housing. Install hub and outer bearing on knuckle housing. It may be necessary to heat bearing cones to 120° C (250° F) before installing. Coat bushings (37) with Loctite 270 and drive into knuckle housing. If bushings (37) are spring pin type, split should be toward direction of rotation or away from direction of rotation. **Split of pin should never be toward center of axle or away from center of axle.** Use four M10 × 100 mm cap screws and nuts to force ring gear on bushings. Secure ring gear with eight new self-locking cap screws (38), tightened to 90 N·m (65 ft.-lbs.) torque. Install thrust washer (34) and snap ring (36) on sun gear (35). Install new "O" ring (27) in groove on hub (25). Install bearings (31) in planetary gears (32) and secure with angular snap rings (30 and 33). Install planetary gears on shafts and secure with snap rings (29).

Before installing planetary carrier, determine correct thickness thrust washer (39) to be used. Measure the distance outer end of sun gear (35) is from flange of hub (25). Next measure distance from flange of planetary carrier (40) to thrust washer seat in carrier. From the difference of the two measurements, deduct 0.3-0.6 mm (0.012-0.024 inch) for the specified free play. The result is the correct thickness thrust washer to be installed. Use grease to stick thrust washer in place and install planetary carrier assembly. Install the two Allen screws (41). Fill hub and planetary to the level plug opening (drain plug positioned so that "OLSTAND" mark is in level position) with SAE 85W-140 GL-5 gear oil. Capacity is 0.75 L (0.79 quart). Install front wheel and tighten nuts to a torque of 300-350 N·m (220-255 ft.-lbs.).

STEERING KNUCKLE HOUSING AND AXLE SHAFT

21. R&R AND OVERHAUL. To remove either steering knuckle housing, first remove wheel hub and planetary as outlined in paragraph 20. Disconnect tie rod from steering knuckle arm. Unbolt and remove upper and lower king pins (15 and 19—Fig. 21). Measure and note thickness and number of shims (16) at upper and lower king pin for aid in reassembly. Carefully remove steering knuckle housing (17). Axle shaft and double "U" joint assembly may be removed with knuckle housing. Lower bearing cone (4), seal cap (5) and "O" ring (6) will fall from axle housing fork (1). Remove upper bearing cone, seal cap and "O" ring.

Bearing cups (3) and grease cups (2) can be removed if necessary.

If desired, axle shaft and "U" joint assembly can be withdrawn for inspection or repair. If renewal is required, oil seal (12) and bushing (13) can be removed from knuckle housing (17) and oil seal (8) and bushing (7) can be removed from axle housing (1). Bushings (7 and 13) should be pressed into position with external groove toward top and internal arrow-shaped grooves pointing toward inside of oil-filled housing (away from seal). Using snap ring pliers, open snap ring (10) and pull axle shaft from "U" joint. Spread snap ring (14) and remove sun gear (35) from "U" joint. "U" joint kits are available and installation procedure is conventional. Be careful not to damage seals (8 and 12) when installing axle and knuckle housing.

Reassemble by reversing the disassembly procedure. Tighten king pin retaining cap screws to a torque of 120 N·m (85 ft.-lbs.). Using a torque wrench on upper king pin cap screw, check rolling drag torque of king pin bearings. Add or remove shims (16), equally at top and bottom, as required to obtain rolling drag torque of 7-8 N·m (5-6 ft.-lbs.). Shims are available in varying thicknesses.

Refer to paragraph 20 when reassembling hub and planetary.

DIFFERENTIAL

NOTE: Although the differential can be removed with right axle housing (3—Fig. 22) attached to tractor, most mechanics prefer removing complete front drive axle assembly from tractor and placing it on stands or a large bench.

22. R&R AND OVERHAUL. To remove the differential, first remove axle assembly as outlined in paragraph 19. Remove both hub and planetary units as in paragraph 20 and steering knuckle housings and axle shafts as in paragraph 21. Remove drain plug (15—Fig. 22) and drain oil from axle housing. Unbolt and remove left axle housing (19). Lift out ring gear and differential unit. Remove bearing cups and shims (17) from each axle housing. Shims (17) are used to adjust backlash of ring gear and pinion and preload of carrier bearings. Measure and record thickness of shims (17) for each side when disassembling to facilitate reassembly and adjustment. Remove "O" ring (18) from housing (19).

To remove the drive pinion (41), unlock lock plate (30) and remove nut (31), washer (32) and drive hub (33). Drive pinion shaft (41) inward using a soft hammer and remove pinion and related parts from axle housing. Remove seal (34) and bearing cone (35). Use a slide hammer puller to remove bearing cup (36). Drive bearing cup (39) out of bore and remove from inside the housing. Measure and note thickness and number of shims (38). Remove crush spacer (37) and use suitable puller to remove bearing cone (40) from drive pinion shaft.

Ring gear (29—Fig. 22) and drive pinion (41) are available only as a matched set. If all original parts are reinstalled, original shims (17L, 17R and 38) can be reused. However, if bearings, differential case, gears or housings are renewed, new shim packs must be determined. Clean and inspect all parts and renew any showing excessive wear or other damage.

Refer to paragraph 22A for disassembly, inspection and repair of self-locking differential used on some models. The following includes service procedures common to both locking differentials or to non-locking types. Application will be noted.

On non-locking differential, use a suitable bearing puller to remove carrier bearing cone (16L—Fig. 22) from ring gear (29). Pry lock plate (28) from ring gear, then remove ten screws (42) attaching ring gear to differential housing (20). Screw three or four M10 × 80 mm screws into differential case in place of removed screws (42), position a disc (such as OTC 27549 or equivalent) on heads of the long screws and use a

suitable bearing puller to push against the disc and heads of screws to separate ring gear (29) from case (20). Remove the puller, disc and the temporarily installed long screws, then remove ring gear. Use a 5 mm ($^{13}\!/_{64}$ inch) diameter punch through one of the threaded holes in case (20) to drive pin (45) out of pinion shaft (22). **DO NOT attempt to drive pin into the threads.** After pinion shaft (22) is withdrawn, side gears (23), pinions (24) and thrust washers (25) can be removed.

Assembly of non-locking differential is reverse of disassembly procedure. Roll pins (43 and 44) should be driven into position with split toward direction of rotation, either front or rear. **DO NOT have split toward center or toward outer diameter of ring gear.** Tighten screws (42) evenly using a crossing pattern to a final torque of 70 N·m (50 ft.-lbs.), then install lock plate (28). Heat bearing cones and inner races for bearings (16L and 16R) to 120° C (250° F) and push into position. Be sure races are seated against shoulders.

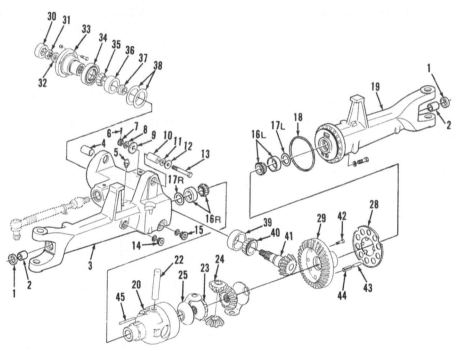

Fig. 22—Exploded view of front drive axle and differential unit used on some models with non-locking differential. Refer to Fig. 23 for self-locking differential.

1. Oil seal	16L. Taper roller bearings	
2. Bushing	16R. Taper roller bearings	32. Washer
3. Axle housing R.H.	17L. Shim	33. Drive hub
4. Bushing	17R. Shim	34. Oil seal
5. Vent	18. "O" ring	35. Bearing cone
6. Cotter pin	19. Axle housing L.H.	36. Bearing cup
7. Slotted nut	20. Differential case	37. Crush spacer
8. Washer	22. Pinion shaft	38. Shims
9. Washer	23. Side gears	39. Bearing cup
10. Pivot tube	24. Bevel pinions	40. Bearing cone
11. Shim	25. Thrust washers	41. Drive pinion
12. Washer	28. Lock plate	42. Special screws
13. Cap screw	29. Ring gear	43. Spring pin (6 × 30 mm)
14. Level plug	30. Lock plate	44. Spring pin (10 × 30 mm)
15. Drain plug	31. Nut	45. Spring pin (5 × 50 mm)

On all models of differential, the correct thickness of shim pack (38) is determined as follows: The measurement from the axle centerline to the bottom of bore in housing for bearing cup (39) in millimeters is stamped, by the manufacturer, on right axle housing just above the axle model plate. Place bearing cone (40) in cup (39) and measure total height of bearing. Add this measurement to dimension stamped in millimeters on face of drive pinion (41). Subtract this sum from dimension stamped in axle housing. The difference is the required thickness of shim pack (38). Place shim pack in bearing bore and install bearing cup (39). Install outer bearing cup (36). Heat bearing cone (40) and install on pinion shaft (41). Slide new crush spacer (37) on pinion shaft, then install pinion shaft in place in axle housing. Heat and install bearing cone (35). Coat lips of seal (34) with grease and install seal in housing bore. Install drive hub (33), washer (32) and nut (31). Tighten nut (31) to 200 N·m (150 ft.-lbs.) torque and measure rolling torque of pinion shaft in bearings. The drive pinion shaft rolling drag torque should be 2-3 N·m (18-26 in.-lbs.) when nut (31) is correctly tightened. Secure nut with lock plate (30).

Adjust ring gear backlash and rolling drag torque of all models as follows: Install one shim (17R—Fig. 22) of 1 mm (0.040 inch) thickness in right axle housing (3). Install bearing cup and make sure cup is seated against shim. Place differential assembly in right axle housing. Using a dial indicator through oil drain hole, check backlash between ring gear and drive pinion. Backlash should be, under this condition (without preload on bearing), 0.2-0.3 mm (0.008-0.012 inch). If necessary, use different thickness shim behind bearing cup to correct backlash.

Place bearing cup on left differential bearing cone and measure distance from right axle housing flange to bearing cup surface as shown in top illustration Fig. 22A. Use spacer blocks and a straightedge to measure distance from left axle housing flange to bearing cup seat as shown in lower illustration in Fig. 22A. Subtract the second measurement from the first measurement. This difference plus 0.2 mm (0.008 inch) will be the correct thickness shim to be used to obtain a rolling drag torque of 1-3 N·m (9-24 in.-lbs.). Install this shim and bearing cup in left axle housing.

Tooth contact pattern can be checked on all models as follows: Lift differential assembly from right axle housing and coat several ring gear teeth with marking blue. Reinstall differential assembly and using four cap screws, attach axle housings together. Turn drive pinion shaft in both directions several turns. Remove left axle housing and lift out differential assembly. Inspect tooth contact pattern and refer to Fig. 24. If ideal contact pattern is not made, recheck all adjustments.

Complete reassembly of all types of differentials as follows: Position differential assembly in right hous-

ing (3—Fig. 22), install new "O" ring (18) in groove of left axle housing (19), then install left axle housing, tightening retaining cap screws to a torque of 190 N·m (140 ft.-lbs.). Reinstall axle shafts and steering knuckle housings as outlined in paragraph 21 and wheel hub and planetary units as in paragraph 20. Reinstall front drive axle assembly as in paragraph 19. Fill wheel hubs and front axle housing to level plug openings with SAE 85W-140 GL-5 gear oil. Capacity for each wheel hub is 0.75 L (0.79 quart) and for axle housings is 4.6-5.0 L (4.9-5.3 quarts). Check level and fill to correct level.

22A. SELF-LOCKING DIFFERENTIAL. Use a suitable bearing puller to remove carrier bearing cone from ring gear (29—Fig. 23). Pry lock plate (28) from ring gear, then remove ten screws (42) attaching ring gear to differential housing (20). Screw three or four M10 × 80 mm screws into differential case in place of removed screws (42), position a disc (such as OTC 27549 or equivalent) on heads of the long screws and use a suitable bearing puller to push against the disc and heads of screws to separate ring gear (29) from case (20). Remove the puller, disc and the temporarily installed long screws, then remove ring gear. Lift out the side gear (23L) and clutch pack (21, 26 and 27) from the open side and keep these parts separate from similar parts for opposite side. Use a 5 mm ($^{13}/_{64}$ inch) diameter punch through one of the threaded holes in case (20) to drive pin (45) out of pinion shaft (22). **DO**

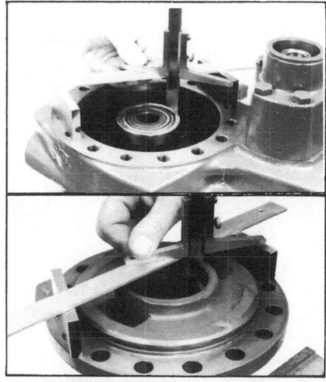

Fig. 22A—Refer to text and these illustrations when checking and adjusting carrier bearing preload.

NOT attempt to drive pin into the threads. After pinion shaft (22) is withdrawn, pinions (24) and thrust washers (25) can be removed. The second side gear (23R) and clutch pack (21, 26 and 27) can then be removed. Although parts (21, 23, 26 and 27) are alike for both sides, parts should be kept together if they are to be reinstalled. The tapered roller bearing cone can be driven from differential case using a punch through interior holes in case.

Assemble differential as follows: If previously used parts are to be reinstalled, they should be located in their original locations or abnormal wear may result. Thickness of clutch packs (21, 26 and 27) should be exactly the same thickness on both sides of pinion gears (24). When measuring thickness of clutch packs and measuring depth of clutch pack in differential case, parts should be **dry.**

Assemble alternating five external lugged plates (26) and four internal splined discs (27), then install one thrust washer (21) and side gear (23R) in differential case. Position thrust washers (25) and pinion gears (24) in case, then insert shaft through case, thrust washers and pinion gears. Use a feeler gage through hole in differential case as shown in Fig. 23A. to measure clearance between first external lugged plate (26) and surface of case. Correct clearance is 0.1-0.2 mm (0.004-0.008 inch). The external lugged plates (26) are available in varying thicknesses and the correct combination of plates should be installed

to provide specified clearance. Be sure that measurement is with dry parts and that external plates (26) and internal splined discs (27) are alternated. When clearance is correct, lubricate parts and install pin (45) to hold pinion shaft. Assemble side gear (23L) and similar clutch pack (21, 26 and 27) in case and measure depth from flange of case to the last external lugged plate (26). Depth measured is the clutch pack operating clearance and should be as near as possible to clearance of pack on other side. Roll pins (43 and 44) should be driven into position with split toward direction of rotation, either front or rear. **DO NOT have split toward center or toward outer diameter of ring gear.** Tighten screws (42) evenly using a crossing pattern to a final torque of 70 N·m (50 ft.-lbs.), then install lock plate (28). Heat bearing cones and inner races for bearings (16L and 16R—Fig. 22) to 120° C (250° F) and push into position. Be sure races are seated against shoulders.

Refer to paragraph 22 for remainder of assembly, adjustment and installation procedures.

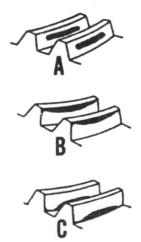

Fig. 23A—Clearance of inner clutch pack can be measured using a feeler gage through hole as shown. Refer to text for procedure.

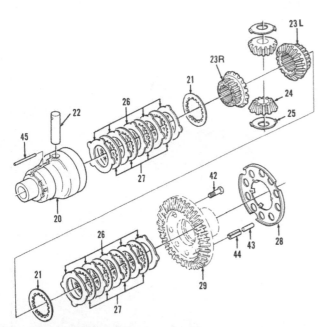

Fig. 23—Exploded view of self-locking front drive differential used on some models. Refer to Fig. 22 for drive pinion shaft, bearings and other parts used.

20.	Differential case		
22.	Pinion shaft	28.	Lock plate
23.	Side gears	29.	Ring gear
24.	Bevel pinions	42.	Special screws
25.	Thrust washers	43.	Spring pin (6 × 30 mm)
26.	External lugged plates	44.	Spring pin (10 × 30 mm)
27.	Internal splined discs	45.	Spring pin (5 × 50 mm)

Fig. 24—Views of typical tooth contact patterns on ring gear.

A. Ideal tooth contact pattern
B. Drive pinion must be moved toward ring gear
C. Drive pinion must be moved away from ring gear

FRONT-WHEEL DRIVE CLUTCH

23. R&R AND OVERHAUL. To remove the front-wheel drive disc clutch assembly, first remove drain plug and drain transmission oil. Disconnect wires from solenoid. Unbolt and remove oil pan and solenoid assembly from clutch housing. Remove the drive shaft as outlined in paragraph 18. Remove plastic sleeve (33—Fig. 26) and washer (32) from front end of shaft (24). Unbolt and remove bearing quill (30) with seal washer (29), oil seal (28), "O" ring (27) and shim (26). Install a cap screw and washer in front end of clutch shaft. Attach a slide hammer puller and pull clutch shaft forward until front bearing cup is free of clutch housing. Support disc clutch assembly with one hand and pull clutch shaft out the front.

NOTE: When clutch shaft is removed, cone for rear bearing (1) and spacer (2) are free to drop out of clutch housing.

Lower disc clutch assembly carefully from clutch housing and catch bearing cone (1) and spacer (2).

To disassemble the clutch, refer to Fig. 26, then unbolt and remove covers (3 and 20). Pull clutch hub (5) from clutch assembly. Tilt clutch assembly and remove thrust washer (16). Use a JDT-24A spring compressor or place assembly in a press and apply pressure to Belleville springs, then remove snap ring (6). Release pressure and remove Belleville springs (7), rear pressure ring (8), internal and external splined discs (9 and 10) and center pressure ring (11). Early models are equipped with a spacer (10A). Remove snap ring (12), spacer ring (13) and front pressure ring (8). Compressed air can be directed through hole in center bore of clutch drum (18) to begin forcing piston (14) with piston ring (15) from bottom of clutch drum. Use screwdrivers to carefully guide piston ring (15) past grooves on clutch drums. Remove front bearing cone (25) and "O" rings (19) from clutch shaft. Remove rear bearing cup (1) from clutch housing, if necessary.

Clean and inspect all parts and renew any showing excessive wear or other damage. Renew all "O" rings and seal rings when reassembling. Press new bushing (4) in clutch hub (5), if necessary.

To reassemble clutch, insert seal ring (17) in groove of clutch drum (18). Install piston ring (15) on piston (14), then install piston in clutch drum. Install front pressure ring (8), release ring (13) and secure with snap ring (12). Rounded edge of snap ring (12) should be toward pressure ring (11). Install center pressure ring (11) with chamfered side facing snap ring. On early models, install one 2.5 mm (0.098 inch) thick external splined disc (10A) as a spacer. Starting with internal splined disc (9), alternately install internal splined and external splined discs (10).

All 2150, all 2155 and early 2350, early 2355, early 2355N, early 2550, early 2555 models are equipped with eight internal splined discs (9) and seven external splined discs (10), plus one 2.5 mm (0.098 inch) spacer (10A). Later 2350, 2355, 2355N, 2550 and 2555 models are equipped with nine internal splined discs (9) and eight external splined discs (10). Later models do not use spacer (10A) or sleeve (22).

On all models, align internal splines of clutch discs (9), position spacer ring (13) over clutch drum, then install rear pressure ring (8) with rounded edge down toward spacer (13). Install the four Belleville springs (7) as shown in Fig. 27. Temporarily install clutch hub and gear to align internal splined discs (9—Fig. 26),

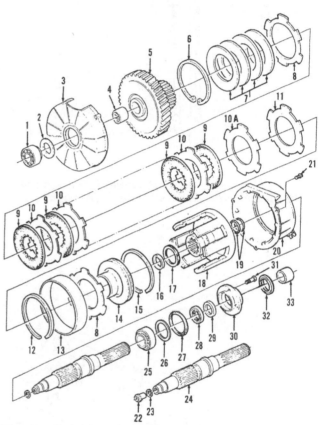

Fig. 26—Exploded view of front-wheel drive clutch used on some models. Refer to text for explanation of key differences.

1.	Taper roller bearing	17.	Seal ring
2.	Spacer	18.	Clutch drum
3.	Cover (rear)	19.	"O" rings
4.	Bushing	20.	Cover (front)
5.	Clutch hub & gear	21.	Cap screw (3)
6.	Snap ring	22.	Sleeve
7.	Belleville springs	23.	Seal ring
8.	Pressure rings	24.	Clutch shaft
9.	Internal splined discs	25.	Front bearing cone & cup
10.	External splined discs	26.	Shim
10A.	Spacer	27.	"O" ring
11.	Pressure ring	28.	Oil seal
12.	Snap ring	29.	Seal washer
13.	Spacer ring	30.	Bearing quill
14.	Piston	31.	Cap screws
15.	Piston ring	32.	Washer
16.	Thrust washer	33.	Sleeve

then use spring compressor JDT-24A or a press, compress Belleville springs and install snap ring (6). Install thrust washer (16) in clutch hub and gear (5) with chamfered side facing clutch drum. Carefully insert clutch hub and gear into clutch discs. If splines

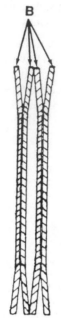

Fig. 27—Sectional view showing correct installation of front-wheel drive clutch Belleville springs (B) used.

of clutch discs (9) are not aligned, it may be necessary to compress Belleville springs (7) again, remove snap ring (6), release springs and again align the splines. When the assembled clutch can be installed fully over hub, install covers (3 and 20) and secure with three cap screws (21).

Before installing clutch assembly, check and adjust, if necessary, clutch shaft axial free play. Install rear bearing cup (1) in clutch housing. Install clutch shaft (24) with rear bearing cone (1), spacer (2) and front bearing (25). Install bearing quill (30) without shims (26) and secure with three cap screws (31). With a dial indicator against front end of clutch shaft, move shaft forward and rearward by hand and check axial end play with a dial indicator. Remove bearing quill (30) and add sufficient thickness of shims to reduce end play of shaft (24) in bearings to 0.02 mm (0.0008 inch) preload to 0.03 mm (0.0012 inch) play.

Remove rear bearing cone (1) and spacer (2) from shaft. Use light grease to stick bearing cone in bearing cup (1) in clutch housing. Install new seal ring (23) on sleeve (22) and new "O" rings (19) in shaft grooves at each side of hydraulic oil port. Shaft (24) is different on later models and sleeve (22) is not used. Hold clutch assembly in clutch housing and insert clutch shaft until spacer (2) can be installed. Install spacer with chamfered side facing bearing (1). Push shaft fully to the rear and install front bearing cup (25).

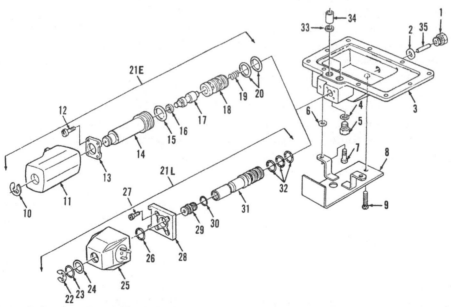

Fig. 28—Exploded view of clutch solenoid and relative parts. Later solenoid/valve assembly is shown at 21L and early type is at 21E.

1. Drain plug	10. "E" ring	19. Spring	27. Allen screw (4)
2. Seal ring	11. Solenoid coil	20. "O" rings	28. Retaining plate
3. Oil pan	12. Allen screw (4)	21E. Solenoid/valve unit	29. Plug
4. "O" ring	13. Retaining plate	21L. Solenoid/valve unit	30. "O" ring
5. Test plug	14. Solenoid core	22. "E" ring	31. Core sleeve
6. Washer	15. "O" ring	23. "O" ring	32. "O" rings
7. Cap screw (2)	16. Back-up washer	24. "O" ring	33. Seal rings (2)
8. Solenoid guard	17. Spool	25. Solenoid coil	34. Tube
9. Cap screw	18. Sleeve	26. "O" ring	35. Pin (8 × 36 mm)

Install the determined shim pack (26) and new "O" ring (27). Install bearing quill (30), with seal washer (29) and oil seal (28), and secure with screws (31).

Reinstall remaining parts by reversing removal procedures. Fill transmission to level mark on dipstick. Install drive shaft as outlined in paragraph 18.

CLUTCH OIL PRESSURE TEST

24. To check front-wheel drive clutch oil pressure, operate tractor until transmission oil temperature is 60° C (140° F). Remove test plug (5—Fig. 28) and install a 2000 kPa (300 psi) test gage.

CAUTION: If the plug has ⁹/₁₆ or ⁵/₈ inch head, the threads should be English; if plug removal requires ¾ inch or metric size wrench, port probably has metric threads. If there is any question, check thread size before attaching test gage.

With engine operating at 1000 rpm and front-wheel drive control switch in "Disengaged" position, operating pressure should be approximately 1050 kPa (150 psi). With control switch in "Engaged" position, pressure gage should read zero. If not, repair solenoid as outlined in the following paragraph.

To check for internal leakage, such as caused by leaking clutch piston seals, install two (matched) 2000 kPa (300 psi) gages as described in Fig. 16. With transmission in "Neutral," dash "MFWD" switch OFF and engine operating at 1000 rpm, pressure at the two test points should be within 170 kPa (25 psi) of the same pressure. If special test cover No. 0745 is used instead of "T" fitting shown in Fig. 16, maximum allowable difference can be 185 kPa (27 psi).

SOLENOID

25. When front-wheel drive control switch is in "Disengaged" position, solenoid is energized causing solenoid spool to direct pressurized oil to the clutch piston. Clutch piston compresses the Belleville springs disengaging the clutch. When control switch is in "Engaged" position, there is no current to the solenoid and a spring forces spool to the off position. In this position, there is no oil pressure at clutch piston and the Belleville springs engage the clutch.

To remove the solenoid, remove drain plug (1—Fig. 28) and drain oil. Unbolt and remove solenoid guard (8). Disconnect wires from solenoid. Remove "E" ring (10 or 22) and pull solenoid coil (11 or 25) from core (14 or 31). Remove Allen screws (12 or 27) and retaining plate (13 or 28). Valve parts can be pulled from bore in pan if cleaning, repair or renewal is required. Early and late parts are different, but operation and service procedures are similar.

Clean and inspect all parts for wear or damage and renew as necessary. Spool and sleeve are available only as a matched set. Use new back-up washer and "O" rings when reassembling and reinstall by reversing the disassembly and removal procedures.

POWER STEERING SYSTEM

A closed center-type power steering system is used on many models and is identified by a mechanical connection from steering shaft arm (Fig. 35) to the steering bellcrank (13—Fig. 1, 13—Fig. 1A, 13—Fig. 7 or 35—Fig. 20A) via a drag link. Pressurized oil is

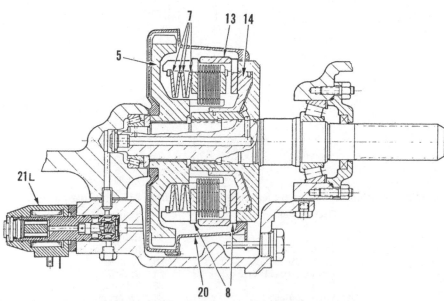

Fig. 28A—Cross section of front wheel drive clutch, cover and valve. Differences may be noted between model shown and some models. Refer to text.

furnished by the main hydraulic pump via a priority-type pressure control valve that is bolted to lower right side of transmission housing. In the event of hydraulic failure or engine stoppage, steering can still be accomplished with the mechanical advantage built into the steering valve assembly. Refer to paragraph 32 and following for service to the hydrostatic steering system used on other models.

TROUBLE-SHOOTING

26. Problems that develop in the power steering system may appear as sluggish steering, loss of power steering, power steering in one direction only or excessive noise in the power steering unit.

1. Sluggish steering can usually be attributed to:
   a. Leakage past valve seats, which usually produces slow steering in one direction only.
   b. Piston sealing ring or "O" ring failure.
   c. Steering valve body (or bodies) leaking.

2. Loss of power steering can usually be attributed to:
   a. Insufficient oil supply from transmission oil pump or main pump.

   b. Pressure control valve out of adjustment.
   c. Clogged hydraulic oil filter.

Chattering system components are usually the result of "O" ring failure in unit or air leak on suction side of pump.

Since the power steering valve is a complete self-contained unit, its only external requirement for proper operation is adequate hydraulic pressure. The transmission pump serves as a charging pump for the main hydraulic pump, which in turn supplies the steering valve via the pressure control valve. The priority type pressure control valve supplies oil to the steering valve first and then to other hydraulic functions.

Pressure control valve output should be tested as outlined in paragraph 29, main hydraulic pump tested as in paragraph 166 and transmission pump tested as in paragraph 124.

Raise front wheels off the ground and with engine stopped, turn steering wheel from one extreme turn position to the other. No binding or hard spots should be encountered. Lower front wheels to the ground, start engine and turn steering wheel from stop to stop. Steering effort should be equal in both direc-

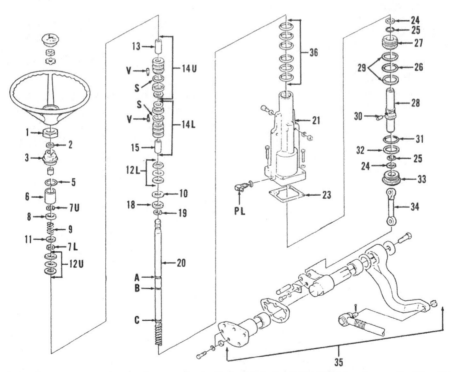

Fig. 29—Exploded view of power steering valve typical of all models. Item (35) is shown in Fig. 34.

PL. Pressure line connector	8. Special washer	19. Snap ring
S. Shims	9. Spring	20. Steering wheel shaft
V. Valve sleeves	10. Shim	21. Housing
1. Jam nut	11. Thrust washer	22. Inlet check valve
2. Oil seal	12L. Lower thrust bearing	23. Gasket
3. Adjuster	12U. Upper thrust bearing	24. Back-up rings
5. "O" ring	13. Sleeve	25. "O" rings
6. Sleeve	14. Valve bodies	26. Seal ring
7L. Snap ring	15. Sleeve	27. Piston
7U. Snap ring	18. Special washer	28. Piston rod

29. Back-up rings
30. Pin
31. "O" ring
32. Back-up ring
33. Piston rod guide
34. Steering rod
35. Steering shaft
36. "O" rings
37. Spacer

tions. Excessive steering effort in one direction only would indicate trouble in steering valve body and/or steering valve for that direction. Excessive steering effort in both directions might indicate trouble with piston "O" ring or seal ring, but also may be caused by excessive weight on front wheels. Internal leakage may be caused by dirty or worn valve seats in valve bodies (14—Fig. 29), or by a valve body shim pack being adjusted incorrectly. If adjuster oil seal (2) is leaking, return oil passage may be obstructed.

STEERING VALVE

27. REMOVE AND REINSTALL. Remove steering wheel and interfering covers from around steering valve assembly. Disconnect hydraulic inlet pressure line (PL—Fig. 29) and remove cover (3—Fig. 34). It may be necessary to detach speed control linkage from cover of some models. Center steering shaft yoke in cover hole by turning steering, then remove cap screw (7), lockwasher (8) and pin retainer (9). Thread a ⅜-inch cap screw into end of pin (6) and remove pin. Steering valve now can be disconnected from dash and clutch housing, and removed as a unit.

Reinstall in reverse order of removal. Coat threads of screw (7) with Loctite 277 and tighten to 15 N·m (11 ft.-lbs.) torque. Start the engine and cycle the system several times to remove any air that may be in system.

28. OVERHAUL. With steering valve removed as outlined in paragraph 27, refer to Fig. 29 and proceed as follows: Remove jam nut (1) and oil seal (2). Use special JDH 41-1B wrench or equivalent to remove adjuster (3), then slide sleeve (6) from housing. Rotate shaft (20) counterclockwise until piston rod (28) is completely free of steering housing. Withdraw steering shaft (20) and steering valve from housing (21). Compress special washer (8) against spring (9), then remove top snap ring (7). Release spring and slide valve parts from shaft.

> **NOTE: Upper and lower valve body parts (14) must not be intermixed. If renewal is necessary, renew as a preadjusted assembly. Special tool, JDG183, is required to determine the correct thickness of shims (S—Fig. 30). Also, lower valve housing has two additional oil passages (Fig. 31).**

Use a special pronged wrench to remove piston (27—Fig. 29) from piston rod (28). Pin (30) must be pressed from piston rod for removal of steering rod (34).

Renew adjuster oil seal (2) and bottom seal in bore. Bushing in adjuster (3) should be renewed if worn excessively. Bushing should be installed until flush with bottom of seal bore in adjuster (3). Inspect all parts for distortion, wear or rough spots on threads and ring lands, and renew as necessary.

Install piston (27) on piston rod (28) with dowel holes in piston to top. Use a special pronged wrench and tighten piston to 340 N·m (250 ft.-lbs.) torque. Assemble all parts on steering wheel shaft (20), beginning with snap ring (19) in groove (A). Be sure large chamfers on thrust bearing races (12L and 12U) are toward valve bodies. Clearance between snap ring (7L) and thrust bearing (12U) should be less than 0.1 mm (0.004 inch). If excessive, shims should be added at (10). Install washer (11), spring (9) and special washer (8). Compress spring and install upper snap ring (7U). Renew all "O" rings, back-up rings and seal rings. Lubricate all parts before assembling. Assem-

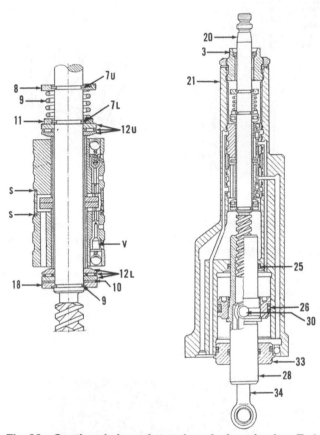

Fig. 30—Sectional view of steering shaft and valve. Refer to Fig. 29 for legend.

Fig. 31—View showing the two additional oil return passages in the lower valve housing.

ble steering rod (34) into piston rod (28) small end first and press pin (30) until flush with outside surface of piston rod. Slip piston rod guide (33) onto piston rod and install assembly into housing. Insert assembly on steering wheel shaft and parts into housing and thread shaft into piston rod. Install sleeve (6) and wrap threads of steering wheel shaft with plastic tape to protect lips of seal (2) as adjuster (3) is screwed into housing. Tighten adjuster to 100 N·m (75 ft.-lbs.) torque. Tighten jam nut (1) to 40 N·m (30 ft.-lbs.) torque.

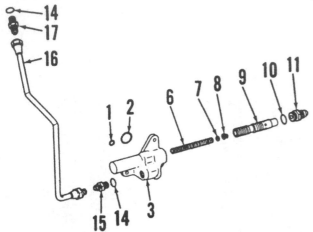

Fig. 32—Exploded view of pressure control valve. Line (16) is pressure line to selective control valve. Orifice (8) is integral with valve (9).

1. "O" ring	
2. "O" ring	10. "O" ring
3. Valve housing	11. Connector
6. Spring	14. "O" ring
7. Shim	15. Connector
8. Orifice	16. Pressure line
9. Control valve	17. Connector

Fig. 33—Connect a 20,000 kPa (3000 psi) test gage to main pump as shown for pressure control valve test.

Reinstall unit on tractor with front wheels in straight ahead position. Install steering rod pin (6—Fig. 34) through yoke (10). Install pin retainer (9), lockwasher (8) and cap screw (7). Coat threads of screw (7) with Loctite 277 and tighten to 15 N·m (11 ft.-lbs.) torque. Install cover (3) with new gasket. Install steering wheel with spoke pointing straight down and tighten nut to 68 N·m (50 ft.-lbs.) torque. Start engine and cycle system several times to purge air from steering valve.

PRESSURE CONTROL VALVE

29. PRESSURE TEST. Remove the ⅜ -inch plug from right side of hydraulic pump, then attach a 20,000 kPa (3000 psi) test gage as shown in Fig. 33. Start engine and operate at approximately 1900 rpm. Check lift system operations and pump standby pressure, which should be 15,170-15,860 kPa (2200-2300 psi). Readjust pump stroke control valve (see paragraph 167) to 10,345 kPa (1500 psi), then attempt to operate rockshaft. The lift system should not function, and if it does, pressure control valve is faulty and should be repaired as in paragraph 30 before proceeding further.

To continue testing pressure control valve, completely lower rockshaft and place rockshaft control lever in raised position. With engine running at 1900 rpm, adjust main hydraulic pump stroke control (raise pressure) until rockshaft raises at its normal rate. This is the regulating point of the pressure control valve and gage should read 11,725-12,410 kPa (1700-1800 psi). If pressure is not as stated, disconnect front oil line, remove fitting (11—Fig. 32) and vary shims (7) as required. Shims are 0.76 mm (0.030 inch) thick and one shim will change pressure 250-280 kPa (35-40 psi).

If pressure control valve cannot be adjusted satisfactorily, remove and service valve as outlined in paragraph 30. Readjust main hydraulic pump stroke control valve to 15,170-15,860 kPa (2200-2300 psi) standby pressure. Remove test gage and reinstall plug.

30. R&R AND OVERHAUL. To remove pressure control valve, drain transmission. Disconnect inlet (front) oil line and if equipped with remote hydraulics, the rear (outlet) oil line. Unbolt and remove unit from transmission housing.

With valve removed, remove fitting (11—Fig. 32), then remove spool (9), shims (7) and spring (6) from housing (3). Retain shims for subsequent installation.

Inspect spool and housing for wear, scoring or other damage. New spool diameters are 19.02-19.05 mm (0.749-0.750 inch) at front and 18.42-18.45 mm (0.725-0.726 inch) at rear. Spring has a free length of 117 mm (4.62 inches) and should test 200-250 N

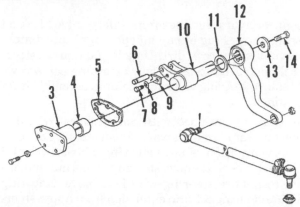

Fig. 34—View of steering shaft component parts which are contained in top of clutch housing.

3.	Cover	9.	Pin retainer
4.	Bushing	10.	Steering shaft
5.	Gasket	11.	Oil seal
6.	Pin	12.	Steering shaft arm
7.	Cap screw	13.	Special washer
8.	Lockwasher	14.	Cap screw

(45-55 lbs.) when compressed to a length of 89 mm (3.5 inches).

When reinstalling, attach oil lines before final tightening of mounting bolts.

STEERING CROSS SHAFT

31. REMOVE AND REINSTALL. The steering cross shaft (10—Fig. 34) can be removed without removing steering valve and steering wheel assembly as follows:

Drain oil from steering shaft housing by removing plug on right side of clutch housing. Remove cover (3), center yoke of steering shaft in cover opening and remove cap screw (7), lockwasher (8) and pin retainer (9). Thread a cap screw into pin (6) and remove pin. Take out button plug on left side of clutch housing for access to cap screw (14). Refer to Fig. 35 and remove

Fig. 35—Remove button plug for access to arm attaching cap screw. Inset shows right side with cover removed.

cap screw and steering shaft arm. Inset in Fig. 35 shows cover removed for access to steering shaft yoke, pin retainer and pin. Turn steering wheel full right, make sure steering shaft yoke is aligned with right cover opening and bump steering shaft out right side of clutch housing. Inspect bushing (4—Fig. 34) in cover (3) and bushing in clutch housing and renew if necessary.

Install by reversing the removal procedure. Coat threads of screw (7) with Loctite 277 and tighten to 15 N·m (11 ft.-lbs.) torque. After cover (3) is installed, tighten arm attaching screw (14) securely, strike arm with a hammer, then tighten cap screw to 350 N·m (260 ft.-lbs.) torque on 2150 and 2255 models; 325 N·m (240 ft.-lbs.) torque on 2155 and 2355N models.

HYDROSTATIC STEERING SYSTEM

A hydrostatic steering system is used on some models and consists of a steering valve assembly, safety valve block, pressure control valve and one or two steering cylinders. In the event of hydraulic failure or engine stoppage, manual steering can be accomplished by means of the geroter pump in the steering valve. Refer to paragraph 26 and following for power steering system used on other models.

TROUBLE-SHOOTING

32. Some of the problems that may occur during operation of power steering and their possible causes are as follows:

1. Steering wheel hard to turn. Could be caused by:

  a. Defective hydraulic pump.
  b. Leaking or missing recirculating ball.
  c. Mechanical parts of front steering system binding.
  d. Ball bearings in steering column damaged.
  e. Leaking steering cylinder.
  f. Valve spool and sleeve binding.

2. Steering wheel turns on its own. Could be caused by:

  a. Leaf springs weak or broken.

3. Steering wheel does not return to neutral position. Could be caused by:
   a. Valve spool and sleeve jammed.
   b. Leakage between valve sleeve and housing.
   c. Dirt or metal chips between valve spool and sleeve.

4. Excessive steering wheel play. Could be caused by:
   a. Inner teeth of rotor or drive shaft teeth worn.
   b. Upper flange of drive shaft worn.
   c. Leaf springs weak or broken.
   d. Drive shaft teeth worn.

5. Steering wheel rotates at steering cylinder stops. Could be caused by:
   a. Excessive leakage in steering cylinder(s).
   b. Rotor and stator excessively worn.
   c. Excessive leakage between valve spool and sleeve and between sleeve and housing.

6. Steering wheel "kicks" violently. Could be caused by:
   a. Incorrect adjustment between drive shaft and rotor.

7. Steering wheel responds too slowly. Could be caused by:
   a. Not enough oil.
   b. Steering control valve worn.

8. Tractor steers in wrong direction. Could be caused by:
   a. Steering cylinder hoses incorrectly connected.
   b. Incorrect timing of drive shaft to rotor.

STEERING VALVE

33. STEERING VALVE LEAKAGE TEST. Operate tractor to warm hydraulic oil to 60°-70° C (140°-160° F). Cycle steering valve lock to lock approximately 15 times to stabilize temperature. Detach both steering cylinder lines from the safety valve ports at the side of control valve. Install caps on lines to prevent entrance of dirt and plug the two steering valve ports with high pressure metal plugs to prevent leakage of hydraulic fluid during test. Operate steering wheel to the left and apply 7 N·m (60 in.-lbs.) torque to steering shaft. Continue turning steering to the left and check number of turns attained in one minute. Repeat this test turning steering to the right. If steering wheel turns more than 1.5 revolutions in one minute, excessive leakage is indicated. Remove steering valve and repair as outlined in the following paragraphs.

34. REMOVE AND REINSTALL. On tractors without Sound Gard Body, remove steering wheel. Swing both dash side panels upward. Identify and disconnect hydraulic lines from safety valve block on steering valve. Plug or cap all openings immediately to prevent entrance of dirt. Unbolt and remove steering column cover. Remove four retaining cap screws and remove steering valve and column assembly from below dash.

On tractors equipped with Sound Gard body, disconnect battery ground straps. Push steering column fully forward, then remove steering wheel cover and the release wheel. Remove steering wheel nut and use puller to remove steering wheel. Remove snap ring and slide steering column extension from upper steering shaft. Remove cover below dash, then unbolt and remove dash side covers. Disconnect all electrical plugs at dash and disconnect shut-off cable. Unbolt and remove dash from panel bracket. Identify and disconnect hydraulic lines from safety valve block on steering valve. Cap or plug all openings immediately to prevent dirt from entering system. Remove steering tilt lever, then unbolt and remove steering column and steering valve assembly.

On all tractors, reinstall steering valve assembly by reversing removal procedure. Start engine and

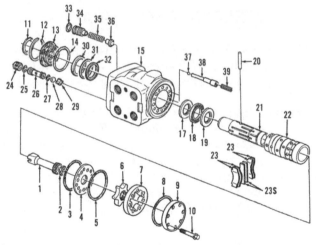

Fig. 38—Exploded view of Char-Lynn hydrostatic steering valve with integral safety shock valves (33-36). Notice that the two center springs (23S) are straight.

1. Drive shaft	22. Valve sleeve
2. Spacer	23. Leaf springs (4)
3. "O" ring	23S. Flat leaf springs (2)
4. Distributor plate	24. Retainer
5. "O" ring	25. "O" ring
6. Rotor	26. Valve seat
7. Stator	27. "O" ring
8. "O" ring	28. Check ball
9. End cover	29. Ball stop
10. Cap screw (7)	30. Back-up ring
11. Oil seal	31. "O" ring
12. Snap ring	32. Seal
13. Gland bushing seal	33. "O" ring
14. "O" ring	34. Adjusting plug
15. Housing	35. Spring
17. Bearing race	36. Shock valve
18. Thrust bearing	37. Suction valve ball
19. Bearing race	38. Suction valve pin
20. Cross pin	39. Spring

operate steering system lock to lock several times to purge air from system.

36. OVERHAUL. Remove steering valve as described in paragraph 34, then thoroughly clean exterior of unit. Unbolt and remove steering column, then place steering valve in a special holding fixture. Remove the two Allen screws and separate safety block (40 or 41—Fig. 39) with intermediate plate (42) and "O" rings (43) from valve of models so equipped. On all models, remove the seven cover retaining screws (10—Fig. 38 or Fig. 39). Remove cover (9), stator (7), rotor (6), spacer ring (2) and "O" rings (5 and 8). Remove distributor plate (4), drive shaft (1) and "O" ring (3). Suction valve parts (37, 38 and 39—Fig. 38) will fall from threaded holes of models so equipped. On all models, hold steering valve vertically and turn valve spool and sleeve to align cross pin (20—Fig. 38 or Fig. 39) parallel to flat side of housing. With cross pin in this position and housing in horizontal position, remove sleeve (22), spool (21), thrust bearing (18) and bearing races (17 and 19) from housing. Remove cross pin (20) from rotary valve and separate spool (21) from sleeve (22). Remove leaf springs (23) from spool. Remove snap ring (12), gland bushing and seal (13) with oil seal (11), "O" ring (14) and quad seal (16—Fig. 39) or seal (30, 31 and 32—Fig. 38) from housing. Unscrew retainer (24—Fig. 38 or Fig. 39) and remove valve seat (26) with "O" rings (25 and 27), check ball (28) and ball stop (29). Safety block valve (33 to 36—Fig. 38) can be removed after unscrewing plug (34) of models so equipped; however, pressure must be checked and adjusted as outlined in paragraph 37 if plug (34) is turned, thus changing adjusted pressure.

Clean and inspect all parts for excessive wear or other damage and renew parts as necessary. Housing (15), spool (21) and sleeve (22) are available only as an assembly. Use all new "O" rings and seals when reassembling. Lubricate all interior parts with clean steering fluid.

Insert spool (21) into sleeve (22) aligning leaf spring slots. Install leaf springs (23), in two sets of two with two flat springs in the middle as shown at (23 and 23S—Fig. 38) of models so equipped or in two sets of three arch to arch as shown at (23—Fig. 39). Use special tool KML 10018-3 when installing leaf springs on all models. Insert cross pin (20—Fig. 38 or Fig. 39) into sleeve and spool. Install "O" ring (14), gland bushing seal (13) with oil seal (11) and snap ring (12). Place bearing race (19), thrust bearing (18), bearing race (17) and quad ring (16—Fig. 39) or seal parts (30, 31 and 32—Fig. 38) on spool. Insert ball stop (29—Fig. 38 or Fig. 39), check ball (28), valve seat (26) with new "O" rings (25 and 27) and retainer (24). Tighten retainer to a torque of 12 N·m (100 in.-lbs.). Lubricate sleeve and install spool and sleeve assembly.

The balance of reassembly is the reverse of disassembly procedure, keeping the following points in mind: When installing rotor (6), make certain that timing dot on end of drive shaft is aligned with a valley on outer side of rotor. Tighten end cover retaining cap screws (10) to a torque of 30-35 N·m (22-25 ft.-lbs.). Be sure that suction valve parts (37, 38 and 39—Fig. 38) are reinstalled in proper holes before installing screws (10). On models so equipped, tighten Allen screws attaching safety valve block (40 or 41—Fig. 39) to 70 N·m (50 ft.-lbs.) torque. If setting of plug (34—Fig. 38) has been disturbed, refer to paragraph 37 for testing and adjusting proper opening pressure.

SAFETY VALVE BLOCK

A safety shock valve is used on all models to act as a relief valve should a front wheel strike a solid object. Piston in steering cylinder is forced to one side and oil pressure increases greatly in hydraulic steering lines. The suction valve on pressure loaded side of safety valve closes immediately. The safety shock valves and suction valves may be located in steering valve housing as shown at (33 to 39—Fig. 38), or in a separate safety valve block (40 or 41—Fig. 39). The specific type used will depend upon model, serial number range and equipment installed on the specific tractor. On all models, the safety valve opens at

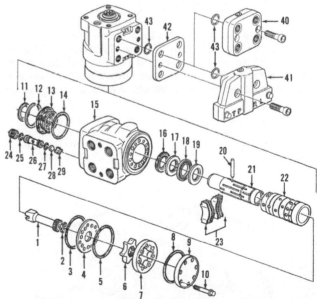

Fig. 39—Exploded view of Char-Lynn hydrostatic steering valve. Notice that all six leaf springs (23) are arched on model shown. If safety shock valve and suction valve are not located in valve housing as shown in Fig. 38, similar valves are located in separate valve block (40 or 41) as shown in Fig. 40 or Fig. 41. Refer to Fig. 38 for legend except the following.

16. Quad ring
40. Safety valve block (Fig. 40)
41. Safety valve block (Fig. 41)
42. Plate
43. "O" rings

20,500 kPa (2980 psi), allowing oil to flow to the low pressure side of steering cylinder. The check valve (2—Fig. 40) prevents oil from returning to hydraulic pump when pump is not operating.

37. TEST AND ADJUST. To test and adjust safety valve block, first remove steering valve assembly as outlined in paragraph 34. Connect a manual hydraulic pump with a 0-34,500 kPa (0-5000 psi) pressure gage to cylinder line port on steering valve (15—Fig. 38) or to appropriate port (3 or 4—Fig. 40 or Fig. 41) of valve block body. Connect a line to return flow port (1) and insert other end into a suitable container. Plug remaining two ports with M18 × 1.5 × 15 mm cap

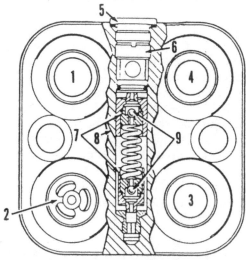

Fig. 40—Cutaway view of safety valve block used on some models equipped with Sound Gard Body. On some models, safety shock valve and suction valves may be located in steering valve body as shown at (33 to 39—Fig. 38) or may use external valve block as shown in Fig. 41.

1. Return flow port
2. Inlet port w/check valve
3. Cylinder line port
4. Cylinder line port
5. End cap
6. Adjusting nut
7. Safety valve
8. Return flow passage
9. Suction valves

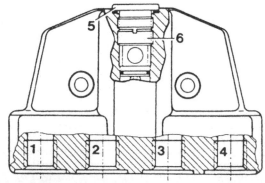

Fig. 41—Safety valve block used on some models not equipped with steering valve with integral safety and suction valves as shown at (33 to 39—Fig. 38). Refer to Fig. 40 for legend.

screws. Operate manual pump until a pressure of 20,500 kPa (2980 psi) is obtained. If pressure then suddenly drops, valve is adjusted correctly.

If necessary, adjust safety valve as follows: Remove end cap and turn adjusting nut (6—Fig. 40 or Fig. 41) or adjusting plug (34—Fig. 38) clockwise to increase pressure or counterclockwise to decrease pressure until correct pressure is obtained.

With pressure correctly adjusted, reinstall steering valve as in paragraph 34.

PRESSURE CONTROL VALVE

A priority type pressure (flow) control valve is located at lower right side of transmission. This valve assures oil supply at sufficient pressure to hydrostatic steering at all times when engine is operating.

38. R&R AND OVERHAUL. To remove the pressure control valve, disconnect hydraulic pump pressure line and line to selective control valve at pressure control valve. Unbolt and remove pressure control valve from transmission case. Remove adapter (14—Fig. 42) with "O" ring (13), then withdraw valve spool (12), orifice (11), shims (10) and spring (9).

Check valve spool (12) and housing (8) for scoring or other damage. Spring free length should be 117 mm (4.62 inches). Spring should test 200-250 N (45-55 lbs.) when compressed to a length of 89 mm (3.5 inches). Check orifice (11) for wear and renew as necessary.

When reassembling, use same number of shims that were removed. Reassemble and reinstall pressure control valve. Check and adjust pressure if necessary as outlined in the following paragraph.

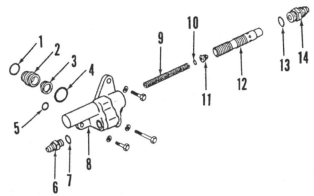

Fig. 42—Exploded view of pressure (flow) control valve used.

1. "O" ring
2. Adapter
3. Back-up ring
4. "O" ring
5. "O" ring
6. Adapter
7. "O" ring
8. Valve housing
9. Spring
10. Shim
11. Orifice
12. Valve spool
13. "O" ring
14. Adapter

39. PRESSURE TEST AND ADJUST. To check pressure control valve, first drain auxiliary reservoir. Remove upper right plug on right side of main pump. Install a 35,000 kPa (5000 psi) test gage in pump. Operate tractor until hydraulic oil is approximately 66° C (150° F). Install a jumper hose across left selective control valve. Set metering valve at maximum and pull left control valve to circulate oil. With engine operating at 800 rpm, test gage should read 11,600-12,300 kPa (1700-1800 psi). If not, stop engine, disconnect pump pressure line and remove adapter (14—Fig. 42). Withdraw valve spool (12) and orifice (11), then add or remove shims (10) as necessary. Adding one shim will increase pressure 250-280 kPa (35-40 psi). Reassemble and retest pressure.

STEERING CYLINDERS

Various types of steering cylinders are used on models with hydrostatic steering. Refer to appropriate following paragraphs for overhaul procedures.

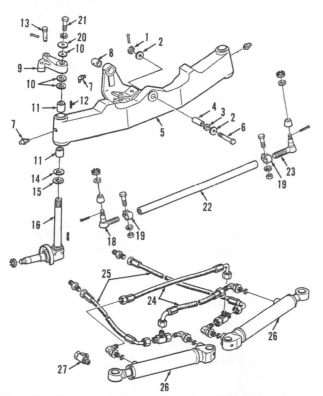

Fig. 45—Exploded view of front axle used on 2355 and 2555 GP models. Two double-acting steering cylinders are used.

1. Slotted nut	10. Washers	19. Clamp
2. Washer	11. Bushings	20. Washer
3. Shim	12. Dowel pin	21. Cap screw
4. Pivot tube	13. Pin	22. Tie rod
5. Axle main member	14. Thrust washer	23. Tie rod end
6. Pivot bolt	15. Washer	24. Hoses
7. Grease fittings	16. Spindle	25. Hoses
8. Bushing	17. Spring pin	26. Steering cylinder
9. Steering arm	18. Tie rod end	27. Test "T" fitting

GP Tractors With Hydrostatic Steering

42. GP tractors with hydrostatic steering are equipped with two double-acting steering cylinders as shown in Fig. 45. Cylinders can be removed after disconnecting hoses and removing cotter pins and retaining pins (13). Cover openings in cylinder to prevent entrance of dirt and plug hoses to reduce leakage and to prevent entry of dirt. To disassemble, remove set screw (42—Fig. 45A), then unscrew rod guide (37). Nut (30) is self locking. "O" ring (35) is installed in seal groove under seal ring (36). Heat seal ring (36) in oil to approximately 55° C (130° F) before installing in groove. Tighten nut (30) to 350 N·m (260 ft.-lbs.) torque.

Models With Two Single-Acting Cylinders

43. On tractors without front-wheel drive and with two single-acting steering cylinders, remove either steering cylinder as follows: Disconnect hydraulic line at steering cylinder. Plug or cap openings to prevent dirt from entering system. Support front of tractor and remove front wheel. Disconnect tie rod from steering arm. Unbolt and remove the axle extension from the axle main member. Detach cylinder ball joint end, unpin rod end and remove cylinder.

To disassemble cylinder, first clean exterior of cylinder and lightly clamp cylinder barrel in a vise.

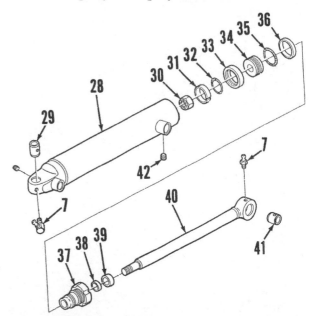

Fig. 45A—Exploded view of double-acting steering cylinder used with axle shown in Fig. 45.

7. Grease fitting		
28. Cylinder		35. "O" ring
29. Bushings		36. Back-up ring
30. Nut		37. Piston rod guide
31. Wear ring		38. Piston rod seal
32. "O" ring		39. Wiper seal
33. Sealing ring		40. Piston rod
34. Piston		41. Bushing

Using a screwdriver through hydraulic port, force snap ring out of groove and off end of piston rod as shown in Fig. 47. Pull piston rod (7—Fig. 46) from cylinder barrel (6). Remove wiper seal (1) and piston rod seal (2) from barrel.

Clean and inspect all parts for excessive wear or other damage and renew as necessary. Install new piston rod seal (2) with lips of seal facing inward, then install new wiper seal (1). Position snap ring (3) in bottom of cylinder barrel. Slide piston rod into cylinder barrel and use screwdriver to install snap ring into groove on piston rod.

Reinstall cylinder by reversing removal procedure. Tighten cylinder retaining slotted nut to a torque of 100 N·m (70 ft.-lbs.) and lock with cotter pin. Tighten axle extension bolts to a torque of 400 N·m (300 ft.-lbs.).

Four-Wheel Drive Models With Hydrostatic Steering

46. The double-acting steering cylinder used on four-wheel drive tractors must be disassembled for

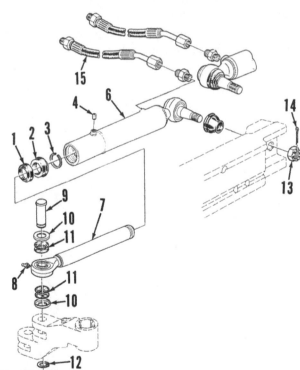

Fig. 46—Exploded view typical of single-acting steering cylinder used on models with adjustable front axle and hydrostatic steering. Two cylinders are used.

1. Wiper seal	9. Pin
2. Piston rod seal	10. Washer
3. Snap ring	11. Seal
4. Bleed plug	12. Snap ring
6. Cylinder barrel	13. Slotted nut
7. Piston rod	14. Cotter pin
8. Grease fitting	15. Hoses

removal. Cylinder casting is an integral part of front drive axle housing.

To disassemble steering cylinder, first remove cotter pin and slotted nut (C—Fig. 20), then disconnect tie rod ends from steering arms. Note position of steering stop clamps (22—Fig. 51), then unbolt and remove stop clamps from each end of piston rod (1). Loosen clamps (A—Fig. 20) and unscrew left and right tie rods (B). Disconnect hydraulic hoses at steering cylinder and cover openings to prevent entrance of dirt. Remove five cap screws (4—Fig. 51); then pull on left end of piston rod (1) to remove all internal parts from left side. Remove inner and outer guides (18 and 3) and shim (2). Pull piston rod and piston assembly out of sleeve (20). Remove "O" rings (19 and 21) from outer diameter of cylinder sleeve. Remove the two scraper rings (13) and seal ring (14) from piston (12). Remove snap rings (7), washers (8), split rings (9), washers (10) and piston (12). Remove "O" ring (11) from piston rod. Remove wiper seal (6) and piston rod seal (5) from outer guide (3). Then, remove wiper seal (6), piston rod seal (5) and "O" ring (17) from inner guide (18).

Clean and inspect all parts for excessive wear, scoring or other damage and renew as necessary. Reassemble cylinder by reversing the disassembly procedure, keeping the following points in mind: Use all new "O" rings, scraper rings and seals when reassembling. Smaller outer diameter of cylinder sleeve (20) must face toward inner rod guide (18). Install outer guide (3) with original shim (2) and tighten cap screws (4) to 70 N·m (50 ft.-lbs.) torque.

NOTE: If inner guide, outer guide or cylinder sleeve are renewed, the following adjustment and selection of shims (2) will be necessary.

Remove outer guide (3) and shim (2). Install a 3 mm (0.118 inch) lead wire between end of sleeve (20) and outer guide (3), without shim. Tighten cap screws (4)

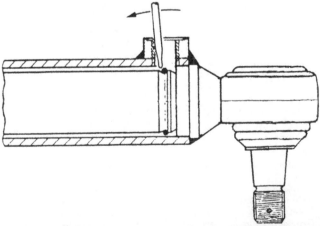

Fig. 47—Using a screwdriver through hydraulic port, force snap ring out of groove and off end of piston rod.

to 70 N•m (50 ft.-lbs.) of torque. Remove cap screws and outer guide and measure thickness of compressed lead wire. Select a shim of this thickness for installation. Actual axial play of sleeve should be 0-0.1 mm (0-0.004 inch). Shims are available in thicknesses from 0.5 mm (0.020 inch) to 1.8 mm (0.070 inch). Reinstall outer guide and selected shim and tighten screws to 70 N•m (50 ft.-lbs.) of torque.

When attaching rod ends to piston rod (1), apply Loctite 242 to threads, tighten rod ends to 250 N•m (180 ft.-lbs.) torque; then lock with lock plate. Tighten slotted nuts of tie rod ends attached to steering knuckles (17—Fig. 21) to 90 N•m (65 ft.-lbs.) torque for M18 slotted nuts; 130 N•m (95 ft.-lbs.) torque for M20 slotted nuts. If necessary, adjust toe-in to 3 mm (⅛ inch). Clamp screws to maintain toe in adjustment should be tightened to 30 N•m (22 ft.-lbs.) torque.

ENGINE AND COMPONENTS

All 2150, 2155 and 2255 models are equipped with 179 cubic inch displacement three cylinder, naturally aspirated, direct injection diesel engines.

All 2350, 2355 and 2550 models are equipped with 239 cubic inch displacement four cylinder, naturally aspirated, direct injection diesel engines.

All 2355N models are equipped with 179 cubic inch displacement three cylinder, turbocharged, direct injection diesel engines.

All 2555 models with collar shift transmissions are equipped with 239 cubic inch displacement four cylinder, naturally aspirated, direct injection engines.

All 2555 models with synchronized shift transmissions are equipped with 239 cubic inch displacement four cylinder, turbocharged, direct injection diesel engines.

All these engines have a bore of 106.5 mm (4.19 inches) and a stroke of 110 mm (4.33 inches) and while similar, differences will be noted.

REMOVE AND REINSTALL

All Models

47. To remove engine and clutch assembly, first drain cooling system and, if engine is to be disassembled, drain oil pan. Disconnect batteries and remove front end weights, if so equipped. Remove side grilles,

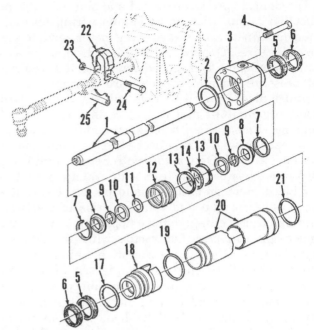

Fig. 51—Exploded view of double-acting steering cylinder components used on four-wheel drive tractors with hydrostatic steering.

1. Piston rod	
2. Shim	13. Scraper rings
3. Outer guide	14. Seal ring
4. Cap screw (5)	17. "O" ring
5. Piston rod seal	18. Inner guide
6. Wiper seal	19. "O" ring
7. Snap rings	20. Sleeve
8. Washers	21. "O" ring
9. Split rings	22. Stop
10. Washers	23. Nut
11. "O" ring	24. Screw
12. Piston	25. Spacer

hood and frame side rails. Disconnect fuel return line and fuel gage wire at fuel tank. Detach wire from air cleaner restriction warning switch and air intake pipe from engine intake manifold. Disconnect hydraulic leak-off line from hydraulic reservoir. Shut off fuel at bottom of tank, then remove fuel line between fuel tank and fuel transfer pump. Disconnect upper and lower radiator hoses and on models where necessary remove radiator brace. Relieve pressure, then disconnect hydraulic pump lines, steering lines and oil cooler return line. On models so equipped, detach steering drag link from bellcrank. Loosen hydraulic pump drive shaft clamp screws so front end and pump can be separated from engine. On models equipped with front-wheel drive, unbolt front drive shaft from drive pinion flange. On tractors with air conditioning, loosen condenser attaching screws and pull condenser out from side. On all models, place wooden blocks between front axle and support to prevent tipping. Support tractor under clutch housing and attach a hoist to front end assembly. Remove cap screws attaching front end to engine, then carefully move front end away from engine.

Disconnect speed control rod and shut-off cable at injection pump. Disconnect engine wiring harness, then disconnect cables at starting motor and alternator as required. On models so equipped, detach speed-hour meter cable from clutch housing and disconnect oil pressure warning switch wire at right rear of cylinder block. On all models without Sound Gard Body, unbolt front half of foot rests from flywheel housing. Remove cap screws attaching cowling to flywheel housing. Disconnect starting fluid line at intake manifold. On tractors with Sound Gard Body, disconnect heater hoses from water pump and cylinder block. Remove batteries and battery boxes. Remove platform mat and center section of platform. Through this opening, remove upper left hex nut and upper right cap screw. On tractors with air conditioning, disconnect the two air conditioning lines at couplings. On all tractors, disconnect hydraulic oil reservoir vent line at transmission shift cover. Attach lifting eyes to cylinder head and attach a hoist to lifting eyes. Unbolt flywheel housing from clutch housing and carefully move engine straight forward away from clutch housing.

Reassemble tractor by reversing the disassembly procedure. Tightening torques are as follows:

Clutch housing to engine . . . . 230 N·m
(170 ft.-lbs.)
Drag link to steering bellcrank . . . . 75 N·m
(55 ft.-lbs.)
Front drive shaft to drive pinion flange . . . . 75 N·m
(55 ft.-lbs.)
Front support to engine block
(⅝ inch screws) . . . . 230 N·m
(170 ft.-lbs.)

Front support to engine block
(⁹⁄₁₆ inch screws) . . . . 180 N·m
(130 ft.-lbs.)
Front support to oil pan . . . . 400 N·m
(300 ft.-lbs.)
Hydraulic pump drive shaft clamp . . . . 50 N·m
(35 ft.-lbs.)
Oil pan to clutch housing . . . . 230 N·m
(170 ft.-lbs.)

CYLINDER HEAD

All Models

48. To remove the cylinder head, drain cooling system and remove grille sides and hood. Disconnect battery ground straps. Detach air intake pipe from intake manifold or turbocharger. Remove cross over tube from turbocharger to intake manifold of models so equipped. Detach or remove muffler and exhaust pipe from all models. Unbolt and remove air intake manifold and exhaust manifold. Remove fuel leak-off line, injector lines and fuel injectors. Remove upper coolant hose and thermostat housing. Disconnect fuel lines and remove fuel filter. On tractors equipped with air conditioning, unbolt and remove compressor. Refrigerant lines need not be disconnected. On all tractors, remove ventilator tube and rocker arm cover. Unbolt and remove rocker arm assembly and lift out push rods. Unbolt and remove cylinder head.

NOTE: Make sure cylinder liners are held down with cap screws and washers if crankshaft is to be turned.

Cylinder head should be flat within 0.102 mm (0.004 inch) when measured across head, measured end to end or diagonally. Cylinder head can be surfaced lightly, but thickness should not be less than 104.11 mm (4.099 inch).

Thoroughly clean gasket surfaces of cylinder head and block. Cylinder liner fire ring of early models is 119 mm (4.685 inch) and later model engines have 120 mm (4.724 inch) diameter fire ring. The late cylinder head gasket can be used with early (119 mm) or the later (120 mm) cylinder liners; however, early gasket for smaller (119 mm) fire ring should be used **ONLY** with cylinder liners with smaller (119 mm) fire ring. Install new head gasket dry, but oil threads of cylinder head retaining cap screws. Use locating dowels in holes (11 and 12—Fig. 52) of three cylinder models; in holes (16 and 17) of four cylinder models to reduce chance of damage to cylinder head gasket. Make sure hardened flat washers are installed on all cylinder head retaining cap screws without flanged head. Cylinder head retaining cap screws can be reused, but the correct type must be installed and

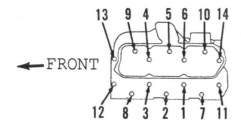

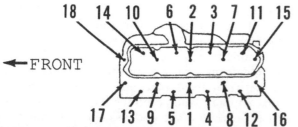

Fig. 52—Tighten cylinder head cap screws evenly in steps using sequence shown. Refer to text.

correctly tightened. Use the sequence shown in Fig. 52 and the following identification and tightening data to tighten all cap screws evenly to the correct torque.

Cap screw part number R53223 is 108 mm (4.25 inch) long and is used with a hardened washer in early 2150, early 2155, early 2255, all 2350 and all 2550 models. The cylinder blocks that use this screw and washer can be identified by a chamfer or 3 mm (0.118 inch) deep counterbore for each of the screws. Tighten screws in the sequence shown in Fig. 52 in the following four steps.

Step 1 . . . . 47 N·m
(35 ft.-lbs.)
Step 2 . . . . 90 N·m
(65 ft.-lbs.)
Step 3 . . . . 130 N·m
(95 ft.-lbs.)
Step 4 . . . . Break engine in for 30 minutes at 1500 rpm, then loosen each cylinder head retaining screw and retorque as in the previous three steps.

Cap screw part number R85363 is 112 mm (4.41 inch) long, has a flanged head and is used **WITHOUT** a hardened washer in later 2150, late 2155, late 2255, all 2355 and all 2555 models. Tighten screws in the sequence shown in Fig. 52 in the following four steps.

Step 1 . . . . 100 N·m
(75 ft.-lbs.)
Step 2 . . . . 150 N·m
(110 ft.-lbs.)
Step 3 . . . . Wait 5 minutes, then recheck to be sure all screws are tightened to the correct torque listed in Step 2.
Step 4 . . . . Turn each screw an additional 50-70 N·m (37-52 ft.-lbs.). Retightening of these screws after engine break-in is not required.

Tighten rocker arm clamp bolts to a torque of 47 N·m (35 ft.-lbs.). Adjust valve clearance as outlined in paragraph 50. Install rocker arm cover and gasket and tighten cover cap screws to 11 N·m (8 ft.-lbs.) torque. On all models, rocker cover gasket must be installed dry. **DO NOT** use liquid gasket or adhesive. Gasket can be reused in most cases if retaining cap screws are not over tightened. Some models use a rocker arm cover made of a composite material that requires additional care when cleaning. Acetone can be used to clean groove of composite cover. Do not over tighten cover retaining screws.

VALVES AND SEATS

All Models

49. The valve seats are integral with the cast iron cylinder head of some models and seat on renewable hardened steel inserts of other models. Replacement seats are available.

Valve face angle is 43½ degrees and seat angle is 45 degrees for all exhaust valves of all models. Face angle for intake valves of naturally aspirated engines is 43½ degrees and seat angle is 45 degrees. Face angle for intake valves of turbocharged models is 29½ degrees and seat angle is 30 degrees. Recommended seat width is 1.5 mm (0.06 inch) for exhaust valves of all models and intake valves of naturally aspirated models. Recommended seat width is 2.0 mm (0.08 inch) for the intake valves of turbocharged models.

Intake and exhaust valve stem diameter is 9.43-9.46 mm (0.371-0.372 inch) with a recommended clearance of 0.05-0.10 mm (0.002-0.004 inch) in guide bores. Valves are available with stem oversizes of 0.08, 0.38 and 0.76 mm (0.003, 0.015, and 0.030 inch) for use if clearance exceeds 0.15 mm (0.006 inch).

With valves installed in cylinder head, check valve recession (distance between valve and cylinder head surface). Valve recession for intake valves should be 0.58-1.19 mm (0.023-0.047 inch). Valve recession for exhaust valves should be 0.97-1.83 mm (0.038-0.072 inch). If recession exceeds 1.63 mm (0.064 inch) for intake valves or 2.26 mm (0.089 inch) for exhaust valves, new valves should be installed and/or seats must be renewed.

Adjust valve clearance using the procedure outlined in paragraph 50.

50. VALVE CLEARANCE. The two-position method of valve adjustment is recommended. Turn engine crankshaft by hand or use a JDE-83 engine rotation tool until "TDC" mark is reached. "TDC" is determined by using timing pin JDE-81-4 or an 8 mm (0.320 inch) rod, 80 mm (3.15 inches) long inserted into hole in left side of flywheel housing. The timing pin will enter hole in flywheel when No. 1 piston is at "TDC." Check the valves of the front No. 1 cylinder to determine whether it's at top of compression stroke or at "TDC" of exhaust stroke. If both valves for the front cylinder are loose, piston is on compression stroke. If the valves are tight, rotate crankshaft one complete revolution and recheck. With No. 1 piston at "TDC" of compression stroke, refer to Fig. 53 or Fig. 53A and adjust the indicated valves. Turn crankshaft one complete revolution until No. 1 cylinder is at "TDC" of exhaust stroke, then adjust remaining valves. Valve clearance should be 0.35 mm (0.014 inch) for intake valves and 0.45 mm (0.018 inch) for exhaust valves.

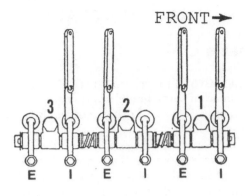

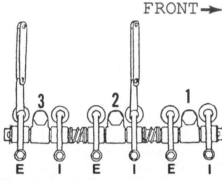

Fig. 53—On three cylinder models, the valves indicated at top can be adjusted when No. 1 piston is at TDC of compression stroke. Turn crankshaft one complete revolution and adjust remaining valves indicated in lower drawing.

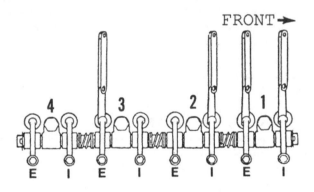

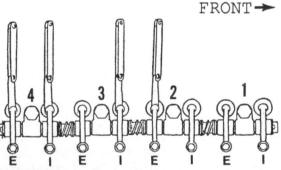

Fig. 53A—On four cylinder models, the valves indicated at top can be adjusted when No. 1 piston is at TDC of compression stroke. Turn crankshaft one complete revolution and adjust remaining valves indicated in lower drawing.

VALVE GUIDES AND SPRINGS

All Models

51. Valve guide bores are an integral part of cylinder head. Valves with 0.08, 0.38 and 0.76 mm (0.003, 0.015 and 0.030 inch) oversize stems are available for service. Guides can also be knurled to reduce clearance with the standard stem. Standard valve guide bore diameter is 9.51-9.53 mm (0.374-0.375 inch) and normal stem clearance is 0.05-0.10 mm (0.002-0.004 inch).

Intake and exhaust valve springs are interchangeable and may be installed either end up. Renew any spring that is distorted, rusted or discolored, or does not conform to the following test specifications:

Free Length . . . . 54 mm
(2.125 inches)
Test at length—
Closed . . . . 240-276 N at 46 mm
(54-62 lbs. at 1.81 in.)
Open . . . . .591-680 N at 34.5 mm
(133-153 lbs. at 1.36 in.)

ROCKER ARMS

All Models

52. The rocker arm shaft is retained to the cylinder head by supports that are attached to the cylinder head by a cap screw through each support. All rocker arms are identical but should not be relocated on shaft after wear pattern has been established. Recommended clearance between rocker arms and shaft is 0.05-0.13 (0.002-0.005 inch). Bushings in rocker arms are not available separately. Rocker arms must be renewed if bushings are excessively worn.

When reassembling, oil lube hole in rocker arm shaft must be facing downward (toward head) and rearward (toward flywheel). Tighten rocker stand attaching cap screws to a torque of 47 N•m (35 ft.-lbs.). Adjust valve clearance as outlined in paragraph 50.

CAM FOLLOWERS

All Models

53. The barrel-type cam followers can be removed from above after removing cylinder head as outlined in paragraph 48. Cam followers operate in unbushed bores in cylinder block and are available in standard size only.

TIMING GEAR COVER AND FRONT OIL SEAL

All Models

54. The front oil seal in the timing gear cover can be removed and a new seal can be installed without removing cover from engine. To gain access to the front of the engine is the same as required to remove the timing gear cover. Seal removal and installation is easier with cover removed.

To remove the timing gear cover, first drain cooling system and disconnect battery cables. Remove front end weights, if so equipped. Remove side grilles and hood. Disconnect fuel return line and fuel gage wire at fuel tank. Detach wire from air cleaner restriction warning switch and air intake pipe from engine intake manifold. Disconnect hydraulic leak-off line from hydraulic reservoir. Shut off fuel at bottom of tank, then remove fuel line between fuel tank and fuel transfer pump. Disconnect upper and lower radiator hoses and on models where necessary, remove radiator brace. Relieve pressure, then disconnect hydraulic pump lines, steering lines and oil cooler return line. On models so equipped, detach steering drag link from bellcrank. Loosen hydraulic pump drive shaft clamp screws so front end can be separated from engine. On models equipped with front-wheel drive, unbolt front drive shaft from drive pinion flange. On tractors with air conditioning, loosen condenser at-taching screws and pull condenser out from side. On all models, place wooden blocks between front axle and support to prevent tipping. Support tractor under clutch housing and attach a hoist to front end assembly. Remove cap screws that attach front end to engine, then carefully move front end away from engine.

Remove fan, fan belt, alternator and water pump. Remove hydraulic pump drive coupling plate (8—Fig. 54) from crankshaft pulley of models without front pto; coupling (6P or 6PH—Fig. 54A) and adapter (8A) from crankshaft pulley of models with front pto. Remove crankshaft pulley retaining cap screw (9—Fig. 54) from models without front pto, attach a suitable puller and remove pulley (12 or 12E). To remove pulley from models with front pto, proceed as follows: Remove pulley retaining screws (12S—Fig. 54A) and bump pulley (12P) toward engine to loosen tapered bore from coupling (12C). Remove screw (9P) and washer (10), then pull collar (12C) and pulley (12P) from end of crankshaft.

To complete removal of the timing gear cover from all models, remove oil pressure regulating plug, spring and valve. Loosen all oil pan attaching cap screws, then unbolt and remove timing gear cover.

The lip-type front oil seal seats on hub of crankshaft pulley (12E—Fig. 54) of early models or a steel wear sleeve (14—Fig. 54 or 54A) pressed on crankshaft of later models. Oil slinger (16) is located on crankshaft with cup side toward front as shown. Clean crankshaft, install new "O" ring (15), coat inner surface of

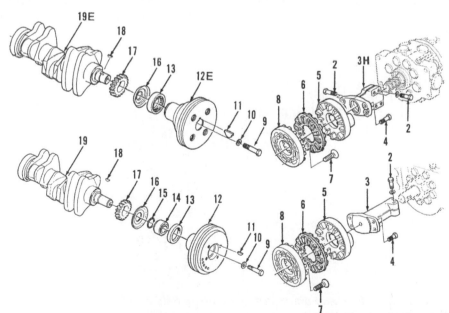

Fig. 54—Exploded view of pump drive and crankshaft used on models without front pto. Note differences for models with standard and high volume hydraulic pumps.

1. Hydraulic pump
2. Clamp screw
3. Pump drive shaft
3H. Drive shaft for high volume pump
4. Cap screws
5. Coupling
6. Cushion
7. Allen screw
8. Adapter coupling
9. Center screw
10. Washer
11. Shaft key
12. Pulley for later models
12E. Pulley for early models
13. Oil seal
14. Seal sleeve
15. "O" ring
16. Oil slinger
17. Crankshaft timing gear
18. Drive key
19. Late crankshaft
19E. Early crankshaft

new wear sleeve with Loctite 609 or equivalent non-hardening sealer and install on crankshaft.

Open side of front oil seal should be toward inside (rear) of front cover. Coat outer edge of seal with Loctite 609 or equivalent non-hardening sealer and drive seal into bore.

When installing timing gear cover, tighten cover retaining cap screws to 50 N·m (35 ft.-lbs.) of torque. Tighten crankshaft pulley or collar retaining cap screw (9) to 150 N·m (110 ft.-lbs.). Tighten pulley to collar screws (12S—Fig. 54A), on models with front pto, evenly to 35 N·m (25 ft.-lbs.) torque. Check alignment of crankshaft pulley after installation, using a dial indicator. Complete the assembly by reversing disassembly procedure.

Tightening torques are as follows:

Drag link to steering bellcrank . . . . 75 N·m
(55 ft.-lbs.)
Front drive shaft to drive pinion flange . . . . 75 N·m
(55 ft.-lbs.)
Front support to engine block
(5⁄8 inch screws) . . . . 230 N·m
(170 ft.-lbs.)
Front support to engine block
(9⁄16 inch screws) . . . . 180 N·m
(130 ft.-lbs.)
Front support to oil pan . . . . 400 N·m
(300 ft.-lbs.)
Hydraulic pump drive shaft clamp . . . . 50 N·m
(35 ft.-lbs.)
Oil pan to clutch housing . . . . 230 N·m
(170 ft.-lbs.)

Oil pan to engine block . . . . 47 N·m
(35 ft.-lbs.)
Oil pan to timing gear cover . . . . 35 N·m
(25 ft.-lbs.)

CAMSHAFT AND TIMING GEARS

All Models

55. The timing gear train (Fig. 55) consists of crankshaft gear, camshaft gear, injection pump gear, oil pump gear, upper idler gear and lower idler gear.

The camshaft must be timed with mark on rim of gear closest to the crankshaft and on a line that passes through the center of camshaft and crankshaft. Crankshaft must be positioned so that No. 1 piston is exactly at "TDC" on compression stroke.

The injection pump gear is marked similarly to the camshaft gear and the mark on the pump gear also should be closest to the crankshaft gear and on a line passing through the center of injection pump and crankshaft. Crankshaft must be positioned so No. 1 piston is at exactly "TDC" on compression stroke.

The easiest way to assure correct timing is to turn crankshaft so that No. 1 piston is at "TDC" of compression stroke. Remove upper idler gear, align camshaft timing mark, align injection pump timing mark, then reinstall upper idler gear without disturbing crankshaft, camshaft or injection pump gears.

56. CAMSHAFT AND GEAR. To remove camshaft, first remove timing gear cover as outlined in

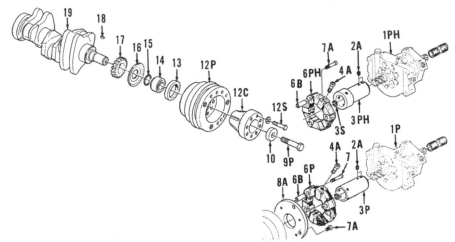

Fig. 54A—Exploded view of pump drive and crankshaft used on models with front pto. Note differences for models with standard and high volume hydraulic pumps.

1P.	Hydraulic pump with through shaft	6PH.	Coupling for high volume pump	12S.	Coupling screws
2A.	Set screws	7.	Allen screws	13.	Oil seal
3P.	Pump drive shaft	7A.	Allen screws	14.	Seal sleeve
3PH.	Drive shaft for high volume pump	8A.	Adapter	15.	"O" ring
3S.	Snap ring	9P.	Center screw	16.	Oil slinger
4A.	Allen screws	10.	Washer	17.	Crankshaft timing gear
6B.	Bushings	12C.	Coupling	18.	Drive key
6P.	Coupling	12P.	Pulley	19.	Crankshaft

paragraph 54 and cylinder head as in paragraph 48. Remove oil pan, cam followers and fuel transfer pump.

Before removing camshaft, mount a dial indicator as shown in Fig. 56 and measure camshaft end play. Recommended end play is 0.05-0.23 mm (0.002-0.009 inch) with a wear limit of 0.38 mm (0.015 inch).

Turn camshaft so thrust plate retaining cap screws are accessible through holes in camshaft gear, then remove the two retaining cap screws. Pull camshaft and thrust plate from engine.

Camshaft operates in unbushed bores in cylinder block. The camshaft gear is keyed and pressed onto front of shaft. Check camshaft and associated parts against these values:

Camshaft journal diameter—
   Desired . . . . 55.87-55.90 mm
                             (2.200-2.201 in.)
   Wear limit . . . . 55.845 mm
                                 (2.199 in.)

Camshaft bearing bores—
   Diameter . . . . 55.98-56.01 mm
                             (2.204-2.205 in.)

Camshaft to bearing diametral clearance—
   Desired . . . . 0.010-0.015 mm
                             (0.004-0.006 in.)
   Wear limit . . . . 0.18 mm
                                 (0.007 in.)

Camshaft thrust plate thickness—
   Desired . . . . 3.96-4.01 mm
                             (0.156-0.158 in.)
   Wear limit . . . . 3.83 mm
                                 (0.151 in.)

Camshaft **intake** lobe height—
   3 cyl. engines before SN. CD666424
     Desired . . . . 6.76-7.26 mm
                             (0.266-0.286 in.)
     Wear limit . . . . 6.50 mm
                              (0.256 in.)
   3 cyl. engines after SN. CD666423
     Desired . . . . 6.93-7.42 mm
                             (0.273-0.292 in.)
     Wear limit . . . . 6.68 mm
                              (0.263 in.)
   4 cyl. engines before SN. CD656457
     Desired . . . . 6.76-7.26 mm
                             (0.266-0.286 in.)
     Wear limit . . . . 6.50 mm
                              (0.256 in.)
   4 cyl. engines after SN. CD656456
     Desired . . . . 6.93-7.42 mm
                             (0.273-0.292 in.)
     Wear limit . . . . 6.68 mm
                              (0.263 in.)

Camshaft **exhaust** lobe height—
   All 3 and 4 cyl. models
     Desired . . . . 6.76-7.26 mm
                             (0.266-0.286 in.)
     Wear limit . . . . 6.50 mm
                              (0.256 in.)

Support camshaft gear and press camshaft from gear if renewal of parts is required. When installing camshaft gear, be sure thrust plate is in position and

Fig. 55—View of timing gear train assembled. Balance shafts and gears (B) are used on four cylinder models.

B. Balance shaft gears
C. Crankshaft gear
G. Camshaft gear
L. Lower idler gear
O. Oil pump gear
P. Injection pump gear
U. Upper idler gear

Fig. 56—Check camshaft end play as shown before removing camshaft.

timing mark is toward front. Reinstall camshaft as follows: The easiest method of aligning timing marks is to remove upper idler gear, turn crankshaft until No. 1 piston is at "TDC" of compression stroke and align camshaft gear and injection pump drive gear timing marks as shown in Figs. 57 and 58. Then, carefully install upper idler gear without moving crankshaft, camshaft or injection pump drive gears.

Tighten camshaft thrust plate cap screws to a torque of 47 N•m (35 ft.-lbs.). Cap screw for upper idler gear should be tightened to 90 N•m (65 ft.-lbs.) torque.

57. IDLER GEARS. Upper and lower idler gears (U and L—Fig. 55) are bushed and operate on station-

Fig. 57—With No. 1 piston at TDC of compression stroke and timing tool (TT) positioned on shafts centerline as shown, camshaft gear timing mark (TM) should be directly under edge of timing tool.

Fig. 58—With No. 1 piston at TDC of compression stroke, injection pump gear (P) is correctly timed if timing mark (TM) is directly below timing tool (TT) when tool is placed between center lines of crankshaft and injection pump shaft.

ary shafts attached to engine front plate with cap screws. Idler gear end play is controlled by thrust washers. Both idler gears are driven by the crankshaft gear (C) and upper idler gear (U) drives camshaft gear (G) and injection pump drive gear (P). Lower idler gear (L) drives oil pump drive gear (O).

To remove both idler gears, remove oil pan and timing gear cover. Remove cap screws and pull gears and thrust washers from shafts. Idler gear shafts can now be removed.

Clean and inspect idler gears and shafts and check against the following specifications:

Shaft OD	44.43-44.46 mm
	(1.749-1.750 in.)
Bushing ID	44.48-44.53 mm
	(1.751-1.753 in.)
Operating clearance	0.02-0.10 mm
	(0.001-0.004 in.)
Max. allowable clearance	0.15 mm
	(0.006 in.)
End play	0.14-0.29 mm
	(0.006-0.012 in.)
Max. allowable end play	0.40 mm
	(0.015 in.)

Reinstall by reversing removal procedure and be sure camshaft, injection pump and crankshaft are timed as indicated in Figs. 57 and 58. Tighten cap screw retaining upper idler gear and shaft to 90 N•m (65 ft.-lbs.) torque. If lower idler gear and shaft are retained by a cap screw, tighten to 130 N•m (95 ft.-lbs.) torque; if retained by a hex nut, tighten nut to 100 N•m (75 ft.-lbs.) torque.

58. CRANKSHAFT GEAR. Renewal of crankshaft gear (C—Fig. 55) requires removal of crankshaft as outlined in paragraph 66. Gear is keyed and pressed on crankshaft. Support crankshaft under first throw when installing new gear. Installation of gear may be eased by heating gear in oil to about 182° C (360° F), but extreme care must be exercised to prevent fire, injury or physical damage. Crankshaft gear has no timing marks. Refer to paragraph 56 for timing camshaft and injection pump to crankshaft.

59. INJECTION PUMP GEAR. To remove the injection pump drive gear, first remove the timing gear cover. The gear is seated on the tapered pump shaft and retained with a hex nut. A Woodruff key in pump shaft assures correct gear to pump shaft timing. Gear shaft is an integral part of pump and gear must be removed for pump removal. Tighten gear retaining nut to 80 N•m (60 ft.-lbs.) of torque.

NOTE: When timing injection pump gear to crankshaft gear, use timing mark indicating number of cylinders of engine. Gear is interchangeable on 3, 4 or 6 cylinder engines.

59A. BALANCE SHAFTS. Four cylinder engines used are equipped with two balance shafts located below the crankshaft on opposite sides of the crankcase. Each shaft is carried in three renewable bushings located in bores in the cylinder block. The right hand balance shaft gear (B—Fig. 55) is driven by the lower idler gear (L) and the left hand balance shaft is driven by the oil pump gear (O). Shafts rotate in opposite directions at twice crankshaft (engine) speed and are designed to dampen the vibration inherent in four cylinder engines.

To remove the balance shafts, first remove the timing gear cover as outlined in paragraph 54. Remove the lower idler gear (L) and oil pump drive gear (O). Identify balance shafts as to right and left. On models with bolt-on balance weights, identify the weights for exact placement (left/right, front/rear) on shafts, then unbolt and remove weights. On all models, unbolt thrust plates and carefully withdraw balance shafts from bores.

Use the following specifications to check balance shafts, bushings and thrust plates.

Shaft journal OD . . . . .38.137-38.163 mm
                                           (1.5014-1.5024 inch)
Bushing ID . . . . .38.177-38.237 mm
                                           (1.5030-1.5054 inch)

Shaft operating clearance—
  Desired . . . . 0.02-0.10 mm
                                           (0.001-0.004 inch)
  Maximum limit . . . . 0.15 mm
                                           (0.006 inch)

Shaft end play—
  Desired . . . . 0.05-0.20 mm
                                           (0.002-0.008 inch)
Thrust plate thickness . . . . .2.97-3.02 mm
                                           (0.117-0.119 inch)

Renew any parts that do not meet specifications. The two front balance shaft bushings can be renewed with engine in tractor; however, the engine, flywheel and flywheel housing must be removed to install rear bushings. Removal of these parts will permit staking the rear bushings.

Use a piloted driver, such as JD-249, to drive front bushings in from front with oil hole aligned with passage in block and with front of bushing flush with chamfer in front of block. Bushings must be staked in position using tool JD-255 or equivalent. Install half round portion of special staking tool in ID of bushing so that the staking ball is in round relief in bushing groove directly opposite oil hole. Locate the square half of special staking tool with correct size dowel engaging the alignment hole in block, then check for correct alignment of cap screw holes in special tool. It may be necessary to turn part of the tool end for end

to correctly align the cap screw holes. Install cap screws, check for correct alignment of ball with relief in bushing, dowel with hole and general alignment, then tighten screws evenly until half round part of tool butts against bushing. This procedure will indent the bushing into the dowel hole and prevent it from turning.

If new gears are pressed on shafts, support shaft in holding fixture, position thrust plate over end of shaft, install drive key, then position gear with timing mark out and press gear on shaft. Press gear onto shaft until clearance between thrust plate and gear is within range of desired shaft end play.

Reinstall balance shafts in reverse of removal procedure; however, set engine at TDC before installing the lower idler gear and oil pump gears. Timing marks must be in line with center line between crankshaft and balance shaft. Special tool JD-254 or equivalent is available for checking alignment of timing marks. Be sure timing marks are aligned after idler gear and oil pump gear are installed.

Tighten thrust plate to block screws to 47 N·m (35 ft.-lbs.) torque and the balance weight attaching nut to 60 N·m (44 ft.-lbs.) torque. If lower idler gear is retained by a cap screw, tighten screw to 130 N·m (95 ft.-lbs.) torque. If lower idler gear is retained by a nut, secure with thread lock and tighten to 100 N·m (75 ft.-lbs.) torque. Tighten oil pump drive gear to 100 N·m (75 ft.-lbs.) torque. Stake nut in three areas around threads with center punch to prevent loosening.

60. TIMING GEAR BACKLASH. Excessive timing gear backlash may be corrected by renewing the gears concerned, or in some instances by renewing idler gear bushing and/or shaft. Refer to the following specifications:

Crankshaft gear to upper idler—
  Desired . . . . 0.07-0.30 mm
                                           (0.003-0.012 in.)
  Max. allowable . . . . 0.4 mm
                                           (0.016 in.)

Camshaft gear to upper idler—
  Desired . . . . 0.07-0.35 mm
                                           (0.003-0.014 in.)
  Max. allowable . . . . 0.5 mm
                                           (0.020 in.)

Injection pump gear to upper idler—
  Desired . . . . 0.07-0.35 mm
                                           (0.003-0.014 in.)
  Max. allowable . . . . 0.5 mm
                                           (0.020 in.)

Crankshaft gear to lower idler—
Desired .... 0.07-0.35 mm
(0.003-0.014 in.)
Max. allowable .... 0.5 mm
(0.020 in.)

Oil pump gear to lower idler—
Desired .... 0.04-0.36 mm
(0.0016-0.015 in.)
Max. allowable .... 0.4 mm
(0.016 in.)

4 Cyl. balance gear to lower idler—
Desired .... 0.05-0.40 mm
(0.002-0.016 in.)
Max. allowable .... 0.51 mm
(0.020 in.)

4 Cyl. balance gear to oil pump gear—
Desired .... 0.05-0.36 mm
(0.002-0.014 in.)
Max. allowable .... 0.51 mm
(0.020 in.)

ROD AND PISTON UNITS

All Models

61. Piston and connecting rod assemblies are removed from above after removing cylinder head and oil pan. Secure cylinder liners (sleeves) in cylinder block using short screws and 1/8-inch thick washers to prevent liners from moving as crankshaft is turned. At least four screw-and-washer combinations should be used on 3 cylinder engines; five should be used on

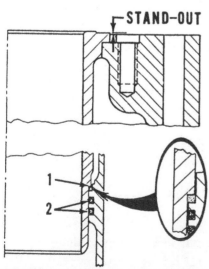

Fig. 59—Sectional view of cylinder sleeve and block showing square sided packing ring (1) and "O" rings (2). Refer to text for correct installation of cylinder sleeves.

4 cylinder engines. Sleeve retaining screws should be tightened to 68 N·m (50 ft.-lbs.) torque.

Pistons are stamped "FRONT," have an arrow on the piston head or are stamped "B" or "H" on underside of piston pin boss. Connecting rods also have the word "FRONT" stamped (embossed) in the web of connecting rod. **Do not stamp head of unmarked pistons for identification before checking underside of piston boss, because upper ring groove may be damaged.** If necessary to mark piston, use a method that does not result in damage to piston. If mark is not visible on piston, combustion bowl of piston should be offset **away from** camshaft side of engine. Connecting rods and pistons may not be originally numbered, but should be stamped with correct cylinder number before removal.

It is suggested that new connecting rod cap retaining screws be used when making final engine assembly. The previously used screws can be installed to check bearing clearance before final assembly. Two different connecting rod cap retaining screws are used and must be installed in proper connecting rod and tightened to proper torque. Phosphate coated cap screws, part number R74194, are 54 mm (2.13 inches) long and should be installed in connecting rods marked by two embedded spot facings on rod cap. Cap screws part number R80033 are 59 mm (2.32 inches) long and should be installed with connecting rod with cap identified by 2 protruding spot facings.

Both types of connecting rod screws should be dipped in clean oil and tightened evenly to 65-75 N·m (50-55 ft.-lbs.) torque upon final assembly.

PISTONS, RINGS AND SLEEVES

All Models

62. Pistons are cam ground, forged aluminum alloy and are fitted with three rings located above the piston pin. Pistons are available in standard size only, but pistons with different ring height and different head height are used for different applications. Be sure correct type is installed for engine and application. Pistons are stamped "FRONT," marked by an arrow on the piston head or stamped "B" or "H" on underside of piston pin boss. **Do not stamp head of unmarked pistons for identification before checking underside of piston boss, because upper ring groove may be damaged.** If necessary to mark piston, use a method that does not result in damage to piston.

The difference in ring groove position can be determined by measuring distance between top of top ring groove and top of piston. Distance on low ring pistons is 18 mm (0.709 inch); 13 mm (0.512 inch) for intermediate ring position pistons; 4 mm (0.158 inch) for pistons with high ring position. Pistons with ring grooves near the top of piston may be safely installed in most models, but all pistons in any engine must be the same. **Be sure to check with manufacturer for correct application.**

Mark "B" indicates standard height from centerline of piston pin to top of piston and "H" mark indicates that head of piston is 0.13 mm (0.0059 inch) higher than standard.

Piston should be assembled to connecting rod with mark ("FRONT," arrow, "B" or "H") on piston and "FRONT" stamped side of connecting rod toward front of engine. If marks on piston are not visible, combustion bowl of piston should be offset **away from** camshaft side of engine.

Top piston ring is of keystone design and a wear gage (JDE-62) should be used for checking piston groove wear. Normal side clearance for middle (2nd) piston ring is 0.038-0.076 mm (0.0015-0.003 inch) with a maximum wear limit of 0.2 mm (0.008 inch). Installation instructions for piston rings are included in ring kits.

The renewable wet-type cylinder sleeves are available in standard size only. Sleeve flange at upper edge is sealed by the head gasket. Sleeves are sealed at lower edge by packing and two "O" rings (Fig. 59). Sleeves normally require loosening, using a sleeve puller, after which they can be withdrawn by hand. Out-of-round or taper should not exceed 0.05 mm (0.002 inch). Check sleeves carefully for rust pits or cracks. If sleeve is to be reused, it should be deglazed, leaving a normal cross-hatch pattern.

When installing sleeves, first make sure sleeve and block bore are absolutely clean and dry. Carefully remove any rust or scale from seating surfaces and from any other areas where loose scale might interfere with sleeve or packing installation. Clean all seating surfaces for packing and "O" rings and grooves in block for "O" rings. If sleeves are being reused, buff rust and scale from outside of sleeve.

Install sleeve without the seals and measure stand-out. See Fig. 59. Check sleeve stand-out at several locations around sleeve flange. Sleeve stand-out should be 0.01-0.10 mm (0.0004-0.004 inch) and should be less than 0.06 mm (0.002 inch) difference between adjacent cylinders. If sleeve stand-out is less than minimum specification, install one special shim between sleeve flange and cylinder block. Shims are available in thicknesses of 0.05 and 0.10 mm (0.002 and 0.004 inch). **Do not install more than one shim.** If stand-out is more than 0.10 mm (0.004 inch), check for scale or burrs under flange. Then, if necessary, select another sleeve. After matching sleeves to all the bores, **stamp the cylinder number location on the top of fire ring at front of sleeve** to facilitate assembly and later identification. **DO NOT stamp lip of sleeve.** Refer to the appropriate following paragraphs for packing and sleeve installation. Refer to preceeding paragraphs for notes on assembling piston to connecting rods. Piston should not

protrude more than 0.33 mm (0.013 inch) higher than face of cylinder block. If protrusion is less than 0.18 mm (0.007 inch) above face of cylinder block with pistons marked "B," pistons marked "H" can be installed.

63. SLEEVE PACKING AND "O" RINGS. Apply liquid soap to the square-sided packing ring (1—Fig. 59) and install over end of cylinder sleeve. Slide packing ring up against shoulder on sleeve. Make sure ring is not twisted and the longer sides are parallel with side of sleeve. Apply liquid soap to "O" rings (2) and install the **black "O" rings in lower grooves** of block; then, install the **red "O" rings in top grooves.** Be sure "O" rings are completely seated in grooves so installing sleeve will not damage "O" rings. Install sleeve carefully into correct cylinder block bores. If sleeve was used previously, be sure installation is in the same cylinder location. If used sleeve is pitted, it is recommended that sleeve be installed 90 degrees from originally installed position so that pitted surface of sleeve is toward front or rear. Stamped cylinder number on top of sleeve fire ring should normally be toward front of engine. Work sleeve gently into position by hand until it is necessary to tap sleeve into position using a hardwood block and hammer.

NOTE: Be careful not to damage packing ring or "O" rings.

Check sleeve stand-out (Fig. 59) with packing installed. The difference between this measured stand-out and similar measurement taken earlier for same sleeve in same bore without packing, will be the compression of the packing. If compression is less than 0.13 mm (0.005 inch), packing ring will not seal properly. Remove sleeve from cylinder block, check packing ring to be sure installation has not cut the packing ring. If shoulders on sleeve and in cylinder block do not provide proper compression of packing ring, install different sleeve and recheck. If a different sleeve will not provide enough compression of packing ring, suggested repair is to install new cylinder block.

63A. SPECIFICATIONS. Specifications of pistons and sleeves are as follows:

Naturally Aspirated Models
Piston skirt*
  diameter . . . . 106.38-106.40 mm
                                                              (4.188-4.189 in.)
Sleeve bore
  diameter . . . . 106.48-106.52 mm
                                                              (4.192-4.194 in.)
Piston skirt* to sleeve clearance—
  Desired . . . . 0.08-0.14 mm
                                                              (0.003-0.005 in.)

Turbocharged Models
Piston skirt*
  diameter . . . . 106.38-106.40 mm
                                            (4.188-4.189 in.)
Sleeve bore
  diameter . . . . 106.48-106.52 mm
                                            (4.192-4.194 in.)
Piston skirt* to sleeve clearance—
  Desired . . . . 0.08-0.15 mm
                                            (0.003-0.006 in.)
* Measured 19 mm (0.740 inch) from bottom of
  skirt, at right angle to piston pin bore.

PISTON PINS

All Models

64. The full floating piston pins are retained in
pistons by snap rings. Pin bushing (5—Fig. 60) is
fitted in upper end of connecting rod and bushing
must be reamed after installation to provide a thumb
press fit for the piston pin. Piston pin to piston recommended clearance is 0.003-0.023 mm (0.0001-
0.0009 inch) for 34.93 mm (1.375 inch) diameter
piston pin; 0.005-0.025 mm (0.0002-0.0009 inch) for
41.28 mm (1.625 inch) diameter pin. Wear limit of pin
is 34.907 mm (1.3743 inch) or 41.257 mm (1.6243
inch).

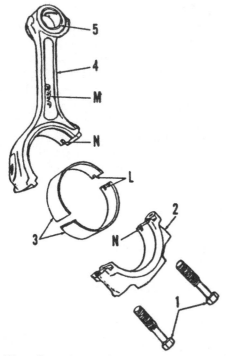

**Fig. 60—The diagonal parting line of connecting rod
should be away from camshaft side.**

1. Cap screws	5. Pin bushing
2. Rod cap	L. Locating tangs
3. Bearing inserts	M. "FRONT" marking
4. Connecting rod	N. Locating notches

Recommended piston pin to bushing clearance is
0.02-0.06 mm (0.0008-0.0024 inch) with a wear limit
of 0.10 mm (0.004 inch) for all models.

CONNECTING RODS AND BEARINGS

All Models

65. The steel-backed aluminum lined bearings can
be renewed from below after removing oil pan and rod
caps. The connecting rod big end parting line is diagonally cut and rod cap is offset away from camshaft
side as shown in Fig. 60. A tongue and groove cap
joint positively locates the cap. Rod marking
"FRONT" should be forward and locating tangs on
bearing inserts must be together when cap is installed.
  Connecting rod bearings are available in undersizes of 0.05, 0.25, 0.51 and 0.76 mm (0.002, 0.010,
0.020 and 0.030 inch) as well as standard size. Refer
to the following specifications:

Crankpin diameter . . . . 69.80-69.82 mm
                                            (2.748-2.749 in.)

Connecting rod bearing clearance—
  Desired . . . . 0.03-0.10 mm
                                            (0.0012-0.004 in.)
  Maximum allowable . . . . 0.16 mm
                                            (0.0062 in.)

Rod bolts —
  Application . . . . Refer to paragraph 61
  Torque . . . . 65-75 N•m
                                            (50-55 ft.-lbs.)

**NOTE: Do not reuse connecting rod cap screws.
Use new cap screws and dip in clean engine oil
when installing. Refer to paragraph 61 for additional notes.**

CRANKSHAFT AND BEARINGS

All Models

66. The crankshaft main bearing inserts can be
renewed after removing oil pan, oil pump and main
bearing caps. All main bearing caps except rear cap
are alike, but should not be interchanged. Main bearing caps should be numbered from front and must be
installed in their original position. Crankshaft end
thrust is controlled by three half thrust washers at
rear main bearing. Two of the thrust washer halves
are located at rear of rear main bearing and one is in
front of rear main bearing. Thrust washers are available in standard and 0.18 mm (0.007 inch) oversize.
Crankshaft end play should be 0.10-0.20 mm (0.004-
0.008 inch) with a maximum end play of 0.38 mm

(0.015 inch). Install main bearing caps with identification marks in numerical order from front to rear and the machined pad in arrow shape pointing to camshaft side of engine. Bearing insert locating tabs should be on same side. Tighten main bearing cap retaining screws to 115 N·m (85 ft.-lbs.) of torque.

To remove the crankshaft, it is necessary to remove engine from tractor. Then, remove oil pan, oil pump, cylinder head and rod and piston units. Remove clutch, flywheel and flywheel housing. Remove timing gear cover, camshaft, injection pump drive gear and both idler gears. Then, unbolt and remove engine front plate from cylinder block.

Check crankshaft and bearings for wear, scoring or out-of-round condition using the following specifications:

Main journal standard diameter .. 79.34-79.36 mm
(3.123-3.124 in.)

Journal taper per 25.4 mm (1 in.) length—
Maximum allowable .... 0.0025 mm
(0.0001 in.)

Journal out-of-round—
Maximum allowable .... 0.08 mm
(0.003 in.)

Main bearing diametral clearance—
Desired .... 0.03-0.10 mm
(0.0012-0.004 in.)
Maximum allowable .... 0.15 mm
(0.006 in.)
Crankpin diameter .... 69.80-69.82 mm
(2.748-2.749 in.)

Connecting rod bearing clearance—
Desired .... 0.03-0.10 mm
(0.0012-0.004 in.)
Maximum allowable .... 0.16 mm
(0.006 in.)

If crankshaft does not meet specifications, either renew it or grind it to the correct undersize. Main bearings are available in standard size and undersizes of 0.05, 0.25, 0.51 and 0.76 mm (0.002, 0.010, 0.020 and 0.030 inch). Be sure to check for local availability before machining crankshaft.

Refer to paragraph 67 for installation of crankshaft rear oil seal and crankshaft wear ring assembly. **DO NOT attempt to reuse old crankshaft rear oil seal and wear ring.**

CRANKSHAFT REAR OIL SEAL

All Models

67. The lip-type rear oil seal is contained in flywheel housing and a wear ring is pressed on crankshaft flange. The seal and wear ring are preassembled and attempts to reinstall seal over wear ring **may** result in oil leakage. A suitable seal installing set, such as JT30040 or KCD1002, should be used to install the seal in flywheel housing and wear ring on crankshaft at the same time.

To renew the seal and wear ring, first remove clutch and flywheel. Seal can be pulled from housing using a jack puller or seal removing tool with tool attached to seal at three equally spaced locations. Wear ring on crankshaft can be removed using a small sharp chisel. Be careful not to damage the crankshaft. Clean crankshaft and inner surface of wear ring with trichloroethylene, then apply a thin coating of Loctite 609 sealer to crankshaft flange. Attach seal installing set pilot to rear of crankshaft, and position seal and wear ring over pilot and on crankshaft. Use seal and pilot installing set to press seal into housing and crankshaft wear ring on to crankshaft simultaneously. Open side of seal and chamfered inner edge of wear ring should both be forward, toward inside of engine. **If seal and wear ring are separated, it is suggested that assembly be discarded, because improper sealing will probably result from attempting to reinstall seal over wear ring, even on new seal.**

FLYWHEEL

All Models

68. To remove flywheel, first remove clutch as outlined in paragraph 102 or 103, then unbolt and remove flywheel from its doweled position on crankshaft. Use two of the flywheel retaining cap screws for jack screws in threaded holes in flywheel and force flywheel off the crankshaft. Use caution to prevent injury or damage when handling the heavy flywheel.

To install a new flywheel ring gear, heat new gear in oil to a temperature of 148° C (300° F). Install ring gear with chamfered end of teeth toward front of flywheel.

Check pilot bearing and renew, if necessary. Install new bearing with shielded side rearward and pack bearing with a high-temperature grease.

When installing flywheel, use new cap screws with Loctite 242 on threads and tighten to 160 N·m (120 ft.-lbs.) of torque. If R74444 flanged head high-strength screws are installed, tighten to 55 N·m (40 ft.-lbs.) torque, then turn screws an additional ⅙ turn (60 degrees).

FLYWHEEL HOUSING

All Models

69. The flywheel housing is secured to rear face of engine block by eight cap screws. Flywheel housing contains the crankshaft rear oil seal and oil pressure sending unit switch. The rear camshaft bore in block is open. It is important that gasket between engine block and flywheel housing is in good condition and cap screws properly tightened to prevent leakage. If flywheel housing has "O" ring grooves, use plain gasket and "O" rings. If housing does not use "O" rings, use gasket with silicone bead. **Do not use adhesives on gasket.** Tighten the ⅜-inch cap screws evenly to 30 N·m (22 ft.-lbs.), then retorque to 47 N·m (35 ft.-lbs.). If so equipped, tighten the four ⅝-inch cap screws to 230 N·m (170 ft.-lbs.) of torque.

Refer to paragraph 67 for installation of new crankshaft rear oil seal. Use of old seal or improper installation procedures will probably result in early seal failure.

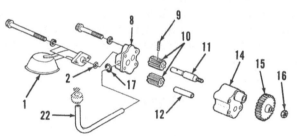

Fig. 61—Exploded view of engine oil pump. Refer to Fig. 64 for oil pressure regulating valve.

1. Oil pick up	12. Idler shaft
2. "O" ring	14. Housing
8. Cover	15. Drive gear
9. Groove pin	16. Nut
10. Pump gears	17. "O" ring
11. Drive shaft	22. Outlet tube

OIL PUMP AND RELIEF VALVE

All Models

70. OIL PUMP. To remove oil pump, first drain and remove oil pan.

On four cylinder models, the left balance shaft is driven by the oil pump gear, so the engine should be set at TDC before removing the pump. Turn engine crankshaft by hand or JDE-83 engine rotation tool until "TDC" mark is reached. "TDC" is determined by using timing pin JDE-81-4 or an 8 mm (0.320 inch) rod, 80 mm (3.15 inches) long inserted into hole in left side of flywheel housing. The timing pin will enter hole in flywheel when No. 1 piston is at "TDC."

On all models, remove nut retaining oil pump drive gear and using a suitable puller, remove gear. Unbolt and remove oil pump.

With pump removed, use Fig. 61 as a guide and proceed as follows: Remove oil pickup (1) and pump cover (8). Remove idler gear and drive gear and shaft (11) from pump housing. Check to see that groove pin (9) is tight in gear and drive shaft. Pin (9) can be renewed if necessary. Check bearing OD of driveshaft (11), which should be 16.02-16.03 mm (0.630-0.631 inch). Check ID of housing bore for drive shaft (11), which should be 16.05-16.08 mm (0.632-0.633 inch). Renew pump if bore diameter exceeds 16.16 mm (0.636 inch).

Idler gear shaft (12) can be pressed from pump housing if renewal is necessary. Diameter of new shaft is 12.32-12.34 mm (0.485-0.486 inch).

Thickness of new gears is 41.15-41.20 mm (1.62-1.622 inches). Install gears and shafts in pump body as shown in Fig. 62 and measure between ends of gear teeth and pump body. This radial clearance should be 0.10-0.16 mm (0.004-0.006 inch) and pump should be renewed if radial clearance exceeds 0.20 mm (0.008 inch). Place a straightedge across pump body as shown in Fig. 63 and measure distance between straightedge and end of pump gears. This axial clear-

Fig. 62—Clearance between gears and housing should be 0.025-0.1 mm (0.001-0.004 inch) when measured as shown.

Fig. 63—Clearance between end of gears and cover should be 0.025-0.15 mm (0.001-0.006 inch) when measured as shown.

ance should be 0.05-0.17 mm (0.002-0.007 inch) and pump should be renewed if axial clearance exceeds 0.22 mm (0.0085 inch).

Reassemble pump by reversing the disassembly procedure. Install new "O" ring in groove of cylinder block before installing pump. On four cylinder models, make sure engine is still at TDC, then install oil pump making sure balance shaft drive gear timing mark is aligned with center line from balance shaft to crankshaft after gears are in place. Check both keyways in balance shafts; the keyways should be at 12 o'clock position.

On all models, tighten screws that attach pump to cylinder block front plate to 47 N·m (35 ft.-lbs.) torque and the pump drive gear retaining nut to 50 N·m (35 ft.-lbs.) torque. Stake nut in three locations around threads with center punch to prevent loosening.

70A. OIL PRESSURE RELIEF VALVE. The oil pressure is controlled by a regulating valve located in the forward end of the cylinder block oil gallery. Relief pressure can be adjusted by adding or removing shims (6—Fig. 64) located between spring (5) and plug (7). With engine at 800 rpm and engine lubricating oil at 90° C (194° F), oil pressure should be 140 kPa (20 psi) minimum for three cylinder models; 100 kPa (14 psi) minimum for four cylinder models. The regulating valve spring should have a free length of approximately 120 mm (4.7 inches) and should test 60-75 N (13.5-16.5 lbs.) when compressed to a length of 42.5 mm (1.68 inches). The plug retaining relief valve assembly in timing gear cover should be tightened to 95 N·m (70 ft.-lbs.) torque.

The relief valve seat (bushing) is pressed into the cylinder block as shown in Fig. 65 and is renewable. When installing new valve seat, use JD-248A or equivalent tool to press seat in block until outer recessed edge is flush with bottom of counterbore. Do not press on, or otherwise damage, the raised inner rim of valve seat.

OIL BYPASS VALVE

All Models

71. The oil bypass valve is located behind the engine front plate next to the oil pressure regulating valve. The bypass valve has a set opening pressure and cannot be adjusted. Bypass spring free length should be 59 mm (2.32 inches) and should test 92-112 N (21-25 lbs.) when compressed to a length of 34 mm (1.34 inches).

OIL FILTER

All Models

72. All models are equipped with a full flow engine oil filter. Manufacturer recommends that oil filter be

Fig. 65—View showing location of engine oil pressure regulating valve seat. Seat is renewable; refer to text.

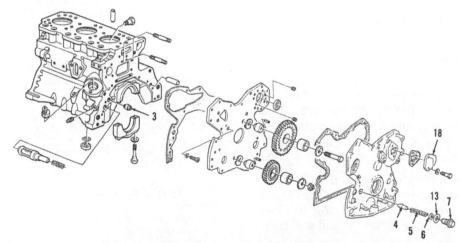

Fig. 64—Exploded view of timing gear cover and front plate showing the engine oil pressure regulating valve.

3. Valve seat (bushing)	5. Spring	7. Plug	18. Injection pump cover
4. Regulating valve	6. Shims	13. Aluminum washer	19. Timing gear opening cover

renewed after each 200 hours of operation. Apply a thin film of oil to filter seal ring, then screw filter on until seal ring touches mounting surface. Tighten filter ¾ to 1¼ turn. Do not overtighten.

ENGINE OIL COOLER

All Models

73. The engine oil cooler is mounted externally on right side of engine block and serves also as the oil filter base. Normal oil cooler servicing consists of cleaning oil passages and cooler bypass valve with suitable solvent.

FUEL SYSTEM

FUEL LIFT PUMP

All Models

74. The fuel lift pump is mounted on right side of cylinder block and is actuated by a lobe on the engine camshaft. The diaphragm-type pump moves fuel from the fuel tank through the fuel filter to the fuel injection pump. Some models of pump can be disassembled and cleaned, then reassembled using new diaphragm, screen, spring and gaskets. Parts of pump body, operating levers and check valves are not available separately for any models. No parts are readily available for some pump models and these pumps should not be disassembled. Renewal of complete pump assembly is required if necessary pump repair parts are not available.

WATER TRAP

Models So Equipped

75. A water trap may be installed in the fuel line between the fuel tank and fuel lift pump in left front of radiator. To drain water and dirt from water trap, loosen drain plug (12—Fig. 66), allow sediment to drain, then tighten drain plug finger tight.

To disassemble water trap for cleaning, shut off fuel at tank and remove cap screw (1) with washer (2) and "O" ring (3). Remove base (10) with seal ring (9), glass bowl (8), housing (7), sediment filter (6) and seal rings (5) from body (4).

Using new seal rings as necessary, reassemble water trap using Fig. 66 as a guide.

FILTER AND BLEEDING

All Models

76. FILTER. All tractors are equipped with a single 2-stage fuel filter. Renew filter element (2—Fig. 67) on Models 2150 and 2255 at every 1000 hours of operation. Renew filter element on Model 2155, 2350, 2355, 2355N, 2550 and 2555 at every 1200 hours of operation. Element should be changed more often if operating under severe conditions or if fuel is contaminated.

To renew the filter element (2), remove drain plug (4), loosen bleed screw (7) and allow filter to drain. Disengage filter element retaining spring (1) and remove element. Make certain filter base (6) is clean, then install new element. Hook top end of retaining

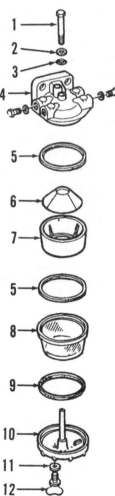

Fig. 66—Exploded view of water trap used on some models.

1. Cap screw		7. Housing
2. Washer		8. Glass bowl
3. "O" ring		9. Seal ring
4. Body		10. Base
5. Seal rings		11. Seal ring
6. Sediment filter		12. Drain plug

spring (1), then hook bottom end. Install drain plug (4). Operate priming lever of fuel lift pump until bubble-free fuel flows from bleed screw, then tighten bleed screw (7).

77. BLEEDING SYSTEM. Whenever fuel system has been run dry, fuel filter changed or a line has been disconnected, air must be bled from fuel system as follows: Make certain there is sufficient fuel in tank and that tank outlet valve is open. On models so equipped, loosen drain plug (12—Fig. 66) on water trap and allow any water or sediment to drain. Tighten drain plug finger tight and loosen bleed screw at top of water trap. When fuel flows from bleed screw, tighten bleed screw. Then, on all models, loosen bleed screw (7—Fig. 67) on fuel filter. Operate priming lever of fuel lift pump until bubble-free fuel flows from bleed screw, then tighten bleed screw. Loosen lower bleed screw on injection pump and operate priming lever of fuel lift pump until fuel flowing from bleed screw is bubble free. Tighten lower bleed screw. Loosen upper bleed screw, pump priming lever until bubble-free fuel flows from bleed screw, then tighten bleed screw. Push priming lever down to lowest position.

NOTE: If no resistance is felt when operating priming lever of fuel lift pump and no fuel is pumped, lift pump arm is on high point of pump cam on camshaft. In this case, turn engine to reposition pump cam to release pump arm.

Loosen pressure line connections at injectors one turn, open throttle and crank engine until fuel flows from loosened connections. Tighten connections and start engine.

INJECTOR NOZZLES

All Models

78. LOCATING A FAULTY NOZZLE. If one engine cylinder is misfiring, it is reasonable to suspect a faulty injector. Generally, a faulty injector can be located by running the engine at low idle speed and loosening, one at a time, each high-pressure line at injector. As in checking spark plugs in a spark ignition engine, the faulty unit is the one that least affects the engine operation when its line is loosened.

If a faulty nozzle is found and considerable time has elapsed since injectors have been serviced, it is recommended that all nozzles be removed and checked, or that new or reconditioned units be installed. Refer to the following paragraphs for removal and test procedures.

79. REMOVE AND REINSTALL. To remove an injector, remove hood and wash injector, lines and surrounding area with clean diesel fuel to remove any accumulation of dirt or other foreign material. Disconnect leak-off line from injector. Disconnect high-pressure line, then cap all openings. Remove cap screw from nozzle clamp and remove clamp and spacer. Pull injector from cylinder head.

NOTE: Unless the carbon stop seal has failed causing injector to stick, injector can be easily removed by hand. If injector cannot be removed by hand, use special puller JDE-38 and pull injector straight out of bore. DO NOT attempt to pry nozzle from its bore.

Before reinstalling injector nozzle, clean nozzle bore in cylinder head using tool JDE-39, then blow out foreign material with compressed air. Turn tool clockwise only when cleaning nozzle bore. Reverse rotation will dull tool. Install new carbon seal (2—Fig. 69) and seal ring (4) whenever injector has been removed. A protector cap and special pilot (Fig. 70) should be used to push new carbon seal onto nozzle body.

NOTE: Nozzle tip may be cleaned of loose or flaky carbon using a brass wire brush. DO NOT use a brush, scraper or other abrasive on Teflon coated surface of nozzle body between the seals. The Teflon coating may become discolored by use, but discoloration is not harmful.

Insert the dry injector in its bore using a twisting motion. Tighten pressure line connection finger tight,

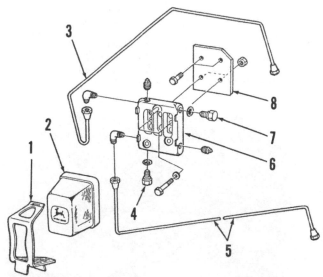

Fig. 67—Exploded view of 2-stage fuel filter typical of all models.

1. Retaining spring	5. Inlet line
2. Filter element	6. Filter base
3. Outlet line	7. Bleed plug
4. Drain plug	8. Mounting bracket

then install hold-down clamp, spacer and cap screw. Tighten cap screw to a torque of 30 N·m (22 ft.-lbs.). Bleed injector if necessary, as outlined in paragraph 77, then tighten pressure line connection only tight enough to prevent leakage. Complete installation by reversing removal procedure.

80. NOZZLE TEST. A complete job of nozzle testing and adjusting requires the use of an approved nozzle tester. Only clean approved testing oil should be used in tester tank. The nozzle should be tested for opening pressure, spray pattern, seat leakage and back leakage. When tested, nozzle should open with a sharp chattering or buzzing sound and cut off quickly at end of injection with a minimum of seat leakage and a controlled amount of back leakage.

Use the tester to check injector as outlined in the following paragraphs:

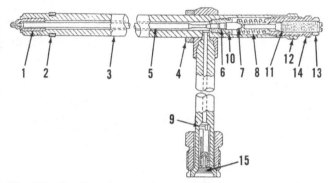

Fig. 69—Sectional view of typical Roosa-Master injector assembly. Nozzle tip (1) and valve guide (6) are parts of finished body and are not serviced separately.

1. Nozzle tip
2. Carbon seal
3. Nozzle body
4. Seal ring
5. Nozzle valve
6. Valve guide
7. Spring seat
8. Pressure spring
9. Edge type filter
10. Seal ring
11. Lift adjusting screw
12. Locknut
13. Locknut
14. Pressure adjusting screw
15. Inlet

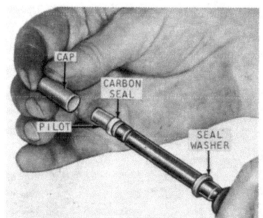

Fig. 70—A protector cap and special pilot should be used when installing new carbon seal.

CAUTION: Fuel leaves the nozzle tip with suffi-cient force to penetrate the skin. Keep unprotected parts of body clear of nozzle spray when testing.

81. OPENING PRESSURE. Before conducting the test, operate tester lever until fuel flows, then attach injector. Close valve to tester gage and pump tester lever a few quick strokes to be sure nozzle valve is not plugged, that spray holes are open and that possibili-ties are good that injector can be returned to service without overhaul.

Open valve to tester gage and operate tester lever slowly while observing gage reading. Opening pres-sure of a used injector should be 20,700 kPa (3000 psi) for naturally aspirated engines; 24,100 kPa (3500 psi) for turbocharged engines. If not correct, adjust open-ing pressure and valve lift as follows:

Loosen locknut (13—Fig. 69) and, while holding pressure adjusting screw (14) from turning, back out lift adjusting screw (11) two turns to ensure against bottoming. Turn pressure adjusting screw (14) until specified opening pressure is obtained. Before tight-ening locknut (13), hold pressure adjusting screw from turning and turn lift adjusting screw (11) inward until it bottoms, then back out ½ turn. Tighten locknut (13) and recheck opening pressure.

NOTE: When adjusting a new injector or an over-hauled injector with a new pressure spring, set opening pressure at 21,700-22,500 kPa (3150-3250 psi) for naturally aspirated engines; 25,100-25,800 kPa (3650-3750 psi) for turbocharged engines. The higher pressure is to allow for initial pressure loss as the spring takes a set.

82. SPRAY PATTERN. The finely atomized nozzle spray should be evenly distributed around the nozzle. Check for clogged or partially clogged orifices or for a wet spray that would indicate a sticking or improp-erly seating nozzle valve. If spray pattern is not satisfactory, disassemble and overhaul injector as outlined in paragraph 85.

83. SEAT LEAKAGE. Pump tester handle slowly to maintain a gage pressure 2800-3500 kPa (400-500 psi) below opening pressure while examining nozzle tip for fuel accumulation. If nozzle is in good condi-tion, there should be no wetness for five seconds. If a drop or undue wetness appears on nozzle tip in five seconds, renew injector or overhaul as in paragraph 85.

84. BACK LEAKAGE. Position nozzle on tester so spray tip is slightly higher than adjusting screw end of nozzle. Then, maintain a gage pressure of 10,300 kPa (1500 psi). After first drop falls from adjusting screw, leakage should be at the rate of 3-10 drops in 30 seconds. Leakage of more than 10 drops indicates excessive wear and injector should be renewed. Leakage of less than three drops indicates dirt and varnish build-up and injector should be overhauled.

85. OVERHAUL. First wash the injector in clean diesel fuel and blow dry with compressed air. Remove carbon seal (2—Fig. 69) and seal ring (4). Clean carbon from spray tip using a brass wire brush. Also clean carbon or other deposits from carbon seal groove in injector body. DO NOT use wire brush or other abrasive on the Teflon coating on outside of nozzle body between the seals. Teflon coating can be cleaned with a soft cloth and solvent. Coating may discolor from use, but discoloration is not harmful.

Place nozzle in a holding fixture and clamp the fixture in a vise. NEVER tighten vise jaws on nozzle body without the fixture. Loosen locknut (13) and back out pressure adjusting screw (14) containing lift adjust screw (11). Remove nozzle body from fixture, invert the body and allow spring (8) and spring seat (7) to fall from nozzle body into your hand. Catch nozzle valve (5) by its stem as it slides from body. If nozzle valve will not slide from body, use special retractor (16481—Fig. 71) to withdraw valve.

Nozzle valve and body are a matched set and should never be intermixed. Keep parts for each injector separate and immerse in clean diesel fuel in a compartmented pan as injector is disassembled.

Clean all parts thoroughly in clean diesel fuel using a brass brush. Hard carbon or varnish can be loosened with a suitable noncorrosive solvent.

NOTE: Never use a steel wire brush or emery cloth on spray tip.

On naturally aspirated engines, clean the four 0.28 mm (0.011 inch) spray tip orifices first with a 0.18-0.21 mm (0.007-0.008 inch) cleaning needle held in a pin vise as shown in Fig. 72. Follow up with a 0.25 mm (0.010 inch) cleaning needle.

On turbocharged engines, clean the four 0.30 mm (0.012 inch) spray tip orifices first with a 0.20-0.23 mm (0.008-0.009 inch) cleaning needle held in a pin vise as shown in Fig. 72. Follow up with a 0.28 mm (0.011 inch) cleaning needle.

Clean valve seat using a Valve Tip Scraper and light pressure while rotating scraper. Use a Sac Hole Drill to remove carbon from sac hole.

Piston area of valve can be lightly polished by hand if necessary, using Roosa-Master No. 16489 lapping compound. Use valve retractor to rotate valve. Move valve in and out slightly while turning, but do not apply down pressure while valve tip is in contact with seat.

Valve and seat are ground to a slight interference angle. Seating areas may be cleaned up if necessary using a small amount of 16489 lapping compound, very light pressure and no more than 3 to 5 turns of

valve on seat. Thoroughly flush all compound from valve body after polishing.

When reassembling, back lift adjusting screw (11—Fig. 69) several turns out of pressure adjusting screw (14), and reverse disassembly procedure using Fig. 69 as a guide. Adjust opening pressure and valve lift as outlined in paragraph 81.

INJECTION PUMP

All Models

86. The Roto-Diesel or CAV/Lucas injection pump is flange mounted on left side of engine front plate and is driven by the upper idler gear of the timing gear train.

Because of the special equipment and specialized training required, service of injection pumps is generally beyond the scope of the average shop. This section will include only the information required for removal, installation and field adjustments of the pump.

87. REMOVE AND REINSTALL. To remove the injection pump, first remove hood, shut off fuel and clean injection pump, lines and surrounding area. Pump can be removed and reinstalled without regard to crankshaft timing position. DO NOT turn crankshaft with injection pump removed.

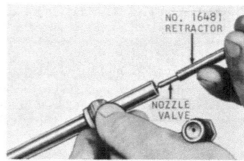

Fig. 71—Use special retractor as shown to remove sticking nozzle valve.

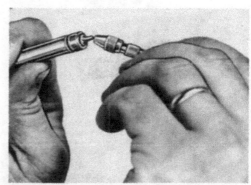

Fig. 72—Using a pin vise and cleaning needle to clean spray tip orifices. Refer to text.

Disconnect or remove fuel inlet, return and pressure lines, throttle rod and stop cable from injection pump. Drain radiator and remove lower radiator hose. Unbolt and remove access plate from front of timing gear cover. Remove drive gear retaining nut and lockwasher. Attach special tool JDG535 or KJD10108 to timing gear cover. Support pump and remove the three mounting stud nuts. Turn center cap screw of special tool until tapered pump shaft is free of drive gear. Remove injection pump, taking care not to lose Woodruff key from pump shaft. Remove special tool.

On all models, pump drive gear will be retained by timing gear cover. DO NOT turn crankshaft with injection pump removed.

To install, turn pump shaft until Woodruff key in shaft aligns with keyway in drive gear. Install lockwasher and nut, then tighten nut to a torque of 80 N·m (60 ft.-lbs.). Tighten pump retaining stud nuts finger tight at this time. Rotate top of pump away from cylinder block as far as slotted holes will allow. Then, rotate pump housing toward block until timing marks (3 and 4—Fig. 74) on engine front plate (1) and pump flange (2) are aligned. Tighten pump mounting stud nuts to a torque of 25 N·m (18 ft.-lbs.). Complete reinstallation by reversing removal procedure. Bleed fuel system as outlined in paragraph 77 and adjust linkage as in paragraph 88.

88. SPEED ADJUSTMENT. Start engine and run until normal operating temperature is reached. Disconnect speed control rod from injection pump and move throttle lever (3—Fig. 75) against high-speed adjusting screw (2) and check engine speed. High idle speed should be within the range listed in the following table. If necessary, turn high-speed adjusting screw in or out as required. Move throttle lever against low idle screw (1). If necessary, adjust low idle screw to obtain recommended engine speed.

High idle speed—
All 2150 models	2610-2660 rpm
2155 models w/collar shift trans.	2610-2660 rpm
2155 models w/synchronized trans.	2410-2510 rpm
All 2255 models	2610-2660 rpm
All 2350 models	2610-2660 rpm
2355 models w/collar shift trans.	2610-2660 rpm
2355 models w/synchronized trans.	2410-2510 rpm
2355N models w/collar shift trans.	2610-2660 rpm
2355N models w/synchronized trans.	2410-2510 rpm
All 2550 models	2610-2660 rpm
2555 models w/collar shift trans.	2610-2660 rpm
2555 models w/synchronized trans.	2410-2510 rpm

Low idle speed—
All 2150 models	700-800 rpm
All 2155 models	750-850 rpm
All 2255 models	700-800 rpm
All 2350 models	700-800 rpm
All 2355 models	750-850 rpm
All 2355N models	750-850 rpm
All 2550 models	700-800 rpm
All 2555 models	750-850 rpm

To adjust shut-off cable, completely push in stop knob and check to be sure stop lever (4) is against stop. Operate engine until warmed up. Pull shut-off knob outward as far as possible and make sure engine stops quickly. Loosen cable clamp and adjust cable if necessary.

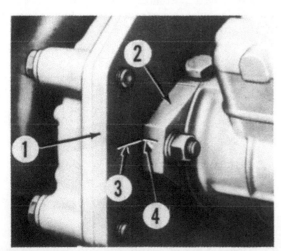

Fig. 74—Installed view of diesel injection pump showing timing marks aligned.

Fig. 75—Installed view of typical diesel injection pump showing control adjusting points.

1. Low idle screw	3. Pump throttle arm
2. High speed screw	4. Stop lever

COOLING SYSTEM

RADIATOR

All Models

89. REMOVE AND REINSTALL. To remove radiator, first drain cooling system, then remove side grille screens and hood. Disconnect fuel return line, hydraulic leak-off line and wiring harness. Remove air intake pipe. Unbolt fan shroud and lay shroud back over fan. On models with hydraulic oil cooler in front of radiator, unbolt cooler and secure to oil reservoir. On models with hydraulic oil cooler beside the radiator, disconnect and plug hoses to and from cooler, plug openings, then unbolt and remover oil cooler. On models equipped with air conditioning, unbolt condenser and remove from side. On all models, disconnect upper and lower radiator hoses and upper radiator brace, then unbolt and remove radiator.

Reinstall by reversing removal procedure. Capacity of cooling system is as follows:

All 2150 models . . . . 10.0 L
(2.65 Gal.)
All 2155 models . . . . 10.0 L
(2.65 Gal.)
All 2255 models . . . . 10.0 L
(2.65 Gal.)
2350 models w/o Sound-Gard body . . . . 13.0 L
(3.4 Gal.)
2350 models w/Sound-Gard body . . . . 15.0 L
(4.0 Gal.)
All 2355 models . . . . 13.0 L
(3.4 Gal.)
All 2355N models . . . . 10.0 L
(2.65 Gal.)
2550 models w/o Sound-Gard body . . . . 13.0 L
(3.4 Gal.)
2550 models w/Sound-Gard body . . . . 15.0 L
(4.0 Gal.)

All 2555 models . . . . 13.0 L
(3.4 Gal.)

Radiator cap has a pressure relief opening set at 100-120 kPa (14-17 psi).

WATER PUMP

All Models

90. REMOVE AND REINSTALL. To remove water pump, first remove radiator as outlined in paragraph 89, then unbolt and remove fan and fan belt. Secure alternator to prevent tipping when brackets are detached, disconnect hoses from water pump, then unbolt and remove water pump from engine.

Reinstall by reversing removal procedure. Adjust fan belt so a 90 N (20 lbs.) pull midway between pulleys will deflect belt 19 mm (¾ inch).

92. OVERHAUL. To disassemble water pump, refer to Fig. 78. Unbolt and remove rear cover (2) and gasket (3). Measure distance from rear face of housing (8) to front flanged surface of pulley or hub (10) and record this distance. **Distance changes the alignment of the drive pulleys and misalignment will cause early belt failure and possible damage.** Use a suitable puller to remove pulley or hub (10) from shaft.

If pump impeller (4) has two tapped holes, use the holes and a suitable puller to remove the impeller, then place water pump in a press with impeller side facing upward. Press shaft and bearing assembly (9) out of impeller and pump body. DO NOT attempt to remove shaft and bearing out impeller side of body because housing bore is stepped. Remove ceramic insert (6) and rubber cup (5) from impeller and pull seal (7) from pump body (8).

If pump impeller (4) does not have puller holes, place water pump in a press with impeller side facing down and supported by housing (8). Press shaft, bearing assembly (9), seal (7) and impeller (4) out of pump

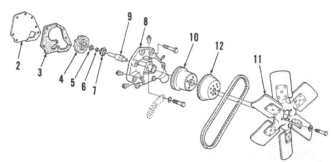

Fig. 77—Exploded view of water pump typical of type used.

2. Cover
3. Gasket
4. Impeller
5. Rubber cup
6. Ceramic insert
7. Seal

8. Pump body
9. Shaft & bearing
10. Pulley or hub
11. Fan blades
12. Pulley sheave

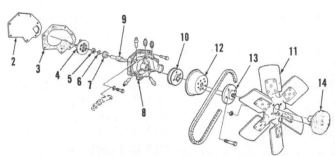

Fig. 78—Exploded view of water pump typical of type used showing installation of fan clutch. Refer to Fig. 77 for legend except the following.

13. Adapter
14. Fan clutch

body (8). If difficulty is encountered, DO NOT attempt to remove shaft and bearing out impeller side of body because housing bore on some models is stepped. Shaft and bearing can be pressed forward out of impeller and housing.

When reassembling, place pump body in a press with front of pump facing upward. Place a washer over pump shaft (use one with a hole large enough so that it contacts only the outer race of bearing and will not damage seal). Press against the washer and outer race of bearing until washer contacts the front of housing. Front of bearing should be flush with front of housing.

Install ceramic insert (6) in rubber cup (5) with "V" groove in insert toward cup. Polished side of ceramic insert should be out toward seal. DO NOT handle ceramic insert with bare hands or seal may leak. Parts must be clean and dry. Coat bore of impeller with a light coat of engine oil, then press cup and insert into impeller until cup bottoms in counterbore. Coat outer edge of seal (7) lightly with oil and install in pump body, using a socket or other driver that contacts only outer flange of seal. Support front end of pump shaft and press impeller (4) on rear of shaft until rear face of impeller is flush with pump body rear face to 0.25 mm (0.010 inch) below flush.

Invert pump assembly and support the unit on rear of pump shaft in a press. Press pulley or hub (10) on front of pump shaft until front face of pulley or hub is the same distance from rear face of pump body as originally measured before disassembly. Complete the assembly by reversing the disassembly procedure. Suggested distance from rear face of pump housing to the front face of pulley or hub is listed in the following table; however, correct distance may be different depending upon equipment and serial number range of tractor. Distance changes the alignment of the drive pulleys and misalignment will cause early belt failure and possible damage.

Pulley flange or hub position—

Model 2150	139 mm
	(5.47 inches)
Model 2155—	
Before Eng. S.N. 743474	139 mm
	(5.47 inches)
After Eng. S.N. 743473	136 mm
	(5.35 inches)
Model 2255	139 mm
	(5.47 inches)
Model 2350	138.94 mm
	(5.47 inches)
Model 2355	138.94 mm
	(5.47 inches)
Model 2355N	154 mm
	(6.02 inches)
Model 2550	138.94 mm
	(5.47 inches)
Model 2555	162.56 mm
	(6.40 inches)

THERMOSTATS

All Models

93. Two thermostats are used on all models. They are located in coolant manifold at front left side of cylinder head. Both thermostats should open at 82° C (180° F). Renew faulty units.

ELECTRICAL SYSTEM

ALTERNATOR AND REGULATOR

All Models

94. OPERATION. All models are equipped with a Bosch 14 volt, 33, 55 or 85 ampere alternator and a regulator that is integral with brush holder.

Model 2150	33A R. Bosch 0120339545
	55A R. Bosch 0120489704
	55A R. Bosch 0120488218
Model 2155	55A R. Bosch 0120489704
	55A R. Bosch 0120488218
Model 2255	33A R. Bosch 0120339545
	55A R. Bosch 0120489704
	55A R. Bosch 0120488218
Model 2350	33A R. Bosch 0120339545
	55A R. Bosch 0120489704
Model 2355	55A R. Bosch 0120489704
	55A R. Bosch 0120488218
	85A R. Bosch 0120484003
Model 2355N	55A R. Bosch 0120489704
	55A R. Bosch 0120488218
Model 2550	33A R. Bosch 0120339545
	55A R. Bosch 0120489704
Model 2555	55A R. Bosch 0120489704
	55A R. Bosch 0120488218
	85A R. Bosch 0120484003

CAUTION: Because certain components of the alternator can be damaged by procedures that will not affect a DC generator, the following precautions must be observed:

a. When installing batteries or connecting a booster battery, the negative post of battery must be grounded.

b. Never short across any terminal of the alternator or regulator.

c. Do not attempt to polarize the alternator.

d. **Disconnect all battery ground straps before removing or installing any electrical unit.**

e. **Do not operate alternator on an open circuit and be sure all leads are properly connected before starting engine.**

Some problems that may occur with the charging system and their possible causes are as follows:

1. Alternator does not charge. Could be caused by:
   a. Slipping fan belt.
   b. Open circuit in charging circuit.
   c. Worn or defective brushes.
   d. Faulty regulator.
   e. Open in rotor field windings.

2. Low or irregular charging output. Could be caused by:
   a. Slipping fan belt.
   b. Intermittent open in charging circuit.
   c. Worn or defective brushes.

3. Excessive charging system voltage. Could be caused by:
   a. Loose alternator connections.
   b. Faulty regulator.

4. Noisy alternator. Could be caused by:
   a. Worn fan belt or misaligned pulley.
   b. Loose pulley.
   c. Worn bearings.
   d. Faulty rectifier diodes.

95. OVERHAUL. To disassemble the alternator, remove brush holder and regulator (12—Fig. 81) and capacitor (13). Immobilize pulley and remove shaft nut, pulley and fan. Mark drive end frame, stator frame and brush end frame for aid in correct reassembly and remove through bolts. Rotor will remain with drive end frame and stator will remain with brush end frame when alternator is disassembled.

Remove the two terminal nuts and three screws securing rectifier (7) to brush end frame and lift out rectifier and stator as a unit. Identify and tag the three stator leads for correct reassembly, then unsolder leads from rectifier diodes using an electric soldering iron and minimum heat. Be careful not to get solder on diode plates or overheat the diodes.

Check brush contact surface of slip rings for burning, scoring or varnish coating. Surfaces must be true to within 0.03 mm (0.0012 inch) and should have diameter of more than 26.8 mm (1.055 inches). Contact surface can be trued by chucking in a lathe. Polish the contact surface after truing using 400 grit polishing cloth until scratches and machine marks are removed. Maximum allowable radial runout of rotor is 0.05 mm (0.002 inch).

Check continuity of rotor windings using an ohmmeter as shown in Figs. 82 and 83. Ohmmeter read-

ing should be 4.0-4.4 between the two slip rings and infinity between either slip ring and rotor pole or shaft.

Stator is "Y" wound, the three individual windings being joined in the middle. Test the windings using an ohmmeter as shown in Fig. 84. Ohmmeter reading should be 0.4-0.44 between any two leads and infinity between any lead and stator frame.

Alternator brushes and shaft bearings are designed for 2000 hours service life. New brushes protrude 10 mm (0.4 inch) beyond brush holder when unit is removed. For maximum service reliability, renew both the brushes and shaft bearings when brushes

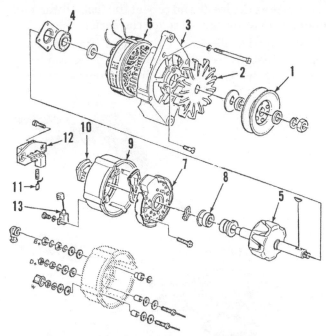

Fig. 81—Exploded view of Bosch 55 ampere alternator. Other Bosch alternators are similar.

1. Pulley
2. Fan
3. Drive end frame
4. Front bearing
5. Rotor
6. Stator
7. Rectifier
8. Rear bearing
9. Brush end frame
10. Insulator
11. Brushes
12. Brush holder & regulator
13. Capacitor

Fig. 82—A reading of 4.0-4.4 ohms should exist between slip rings (1 & 2) when checked with an ohmmeter.

are worn to within 5 mm (0.2 inch) of holder. Solder copper leads to allow for 10 mm (0.4 inch) protrusion using rosin core solder only, and making sure the solder does not seep into and stiffen the wire lead.

The rectifier is furnished as a complete assembly and diodes are not serviced separately. Rectifier unit contains three positive diodes, three negative diodes and three exciter diodes that energize rotor coils before engine is started. If any diodes fail, rectifier must be renewed.

To test positive diodes, touch positive ohmmeter probe to positive heat sink as shown in Fig. 85, and touch negative test probe to each diode lead in turn. Ohmmeter should read at or near infinity for each test. Reverse the leads and repeat the test. Ohmmeter should read at or near zero for the series.

Test negative diodes as shown in Fig. 86. Place negative test probe on negative heat sink and touch each diode lead in turn with positive test probe.

Ohmmeter should read at or near infinity for the series. Reverse test leads and repeat the test. Ohmmeter should read at or near zero for the series.

Test exciter diodes by using the D+ terminal as the base as shown in Fig. 87. Ohmmeter should read at or near infinity with positive test probe on terminal screw and at or near zero with negative test probe touching screw.

Reassemble alternator by reversing the disassembly procedure. Tighten through-bolts to a torque of 4-5.5 N·m (33-48 in.-lbs.) and pulley nut to 35-65 N·m (25-32 ft.-lbs.) torque.

STARTING MOTOR

All Models

96. Bosch starting motors are used. Centers in ends of armature shaft may be slightly off-center. Support armature (12—Fig. 88) by outer diameter of shaft ends when checking commutator and armature plate run-out. Original application and test specifications are as follows:

Original Application

Model 2150	R. Bosch 0001359090
	R. Bosch 0001362312
	R. Bosch 0001362316
Model 2155	R. Bosch 0001362316
Model 2255	R. Bosch 0001359090
	R. Bosch 0001362312
	R. Bosch 0001362316

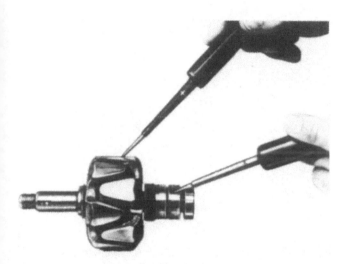

Fig. 83—No continuity should exist between either slip ring and any part of rotor frame.

Fig. 84—No continuity should exist between any stator lead and stator frame.

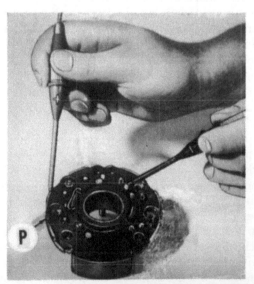

Fig. 85—A near infinity reading should be obtained when positive probe rests on positive heat sink (P) and negative probe touches diode leads as shown. Reverse the probes and reading should be near zero ohms.

Model 2350 . . . . R. Bosch 0001359090
R. Bosch 0001362312
Model 2355 . . . . R. Bosch 0001362312
Model 2355N . . . . R. Bosch 0001362316
Model 2550 . . . . R. Bosch 0001359090
R. Bosch 0001362312
Model 2555 . . . . R. Bosch 0001362312

Robert Bosch Test Specifications
R. Bosch 0001362312
Commutator—
  Minimum diameter . . . . 42.5 mm
    (1.67 inches)
  Max. out of round . . . . 0.03 mm
    (0.0012 inch)
  Insulation undercut . . . . 0.5-0.8 mm
    (0.02-0.03 inch)
Armature—
  Plate out of round, max. . . . . 0.05 mm
    (0.002 inch)
  End play . . . . 0.1-0.3 mm
    (0.003-0.012 inch)
Brush length min. 7.5 mm
    (0.30 inch)

R. Bosch 0001359090
Commutator—
  Minimum diameter . . . . 39.5 mm
    (1.555 inches)
  Max. out of round . . . . 0.03 mm
    (0.0012 inch)
  Insulation undercut . . . . 0.5-0.8 mm
    (0.02-0.03 inch)
Armature—
  Plate out of round, max . . . . 0.05 mm
    (0.002 inch)
  End play . . . . 0.1-0.3 mm
    (0.004-0.012 inch)

Brush length min. . . . . 16 mm
    (0.63 inch)
R. Bosch 0001362316
Commutator—
  Minimum diameter . . . . 42.5 mm
    (1.67 inches)
  Max. out of round . . . . 0.03 mm
    (0.0012 inch)

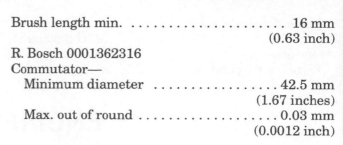

Fig. 87—D+ terminal is used to test exciter diodes. Refer to text.

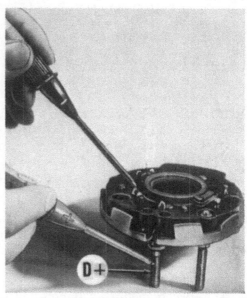

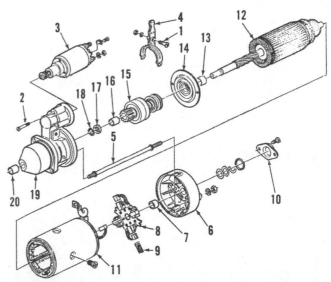

Fig. 88—Exploded view of typical Bosch starter assembly.

1. Screw
2. Screw
3. Solenoid
4. Lever
5. Studs
6. Commutator end housing
7. Bushing
8. Brush holder
9. Springs
10. End cover
11. Field windings assy.
12. Armature
13. Bushing
14. Quill
15. Starter pinion & clutch
16. Bushing
17. Retainer
18. Snap ring
19. Starter nose
20. Bushing

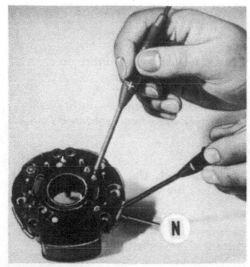

Fig. 86—With negative probe on negative heat sink (N) and positive probe touching diode leads, an infinity reading should be obtained. Reverse the probes and reading should be near zero ohms.

Insulation undercut . . . . 0.5-0.8 mm
(0.02-0.03 inch)
Armature—
Plate out of round, max. . . . . 0.05 mm
(0.002 inch)

End play . . . . 0.1-0.3 mm
(0.003-0.012 inch)
Brush length min. 7.5 mm
(0.30 inch)

ENGINE CLUTCH

ADJUSTMENT

Mechanical Linkage With Reverser

97. PEDAL FREE TRAVEL. When clutch pedal is pushed part way down, clutch control valve lowers reverser system pressure, stopping tractor motion. When pedal is depressed fully, flywheel clutch is released stopping power take off and reverser drive shaft.

To adjust the mechanical linkage, proceed as follows. With engine not running, depress clutch pedal until throwout bearing just contacts clutch release levers, pedal adjusting rod (R—Fig. 89) should be at rear of slot in clutch shaft release arm (C). Measure distance (A) between mounting flange and clutch pedal. Distance (A) should be 133 mm (5¼ inches) for all models and is adjusted by threading yoke (Y) as necessary on rod (R).

With tractor in forward gear and moving, clutch pedal should be pushed 19-32 mm (¾ to 1¼ inches) before any drop occurs in clutch apply pressure. As pedal is depressed further, tractor motion should stop. If adjustment is questioned, check pressure as outlined in paragraph 111.

Other Models With Mechanical Linkage

97A. PEDAL FREE TRAVEL. Clutch pedal free travel (A—Fig. 89A) should be 25 mm (1 inch) and should be readjusted when free travel decreases to 13 mm (½ inch). Adjustment is made by changing the length of the clutch operating rod. Loosen locknut, disconnect one end of rod, then thread rod (R) in or out of yoke (Y) as required. Reconnect rod, tighten locknut, then recheck free travel. If clutch disk is new, adjust free travel (A) to 35 mm (1⅜ inch).

97B. RELEASE LEVERS (DUAL CLUTCH). The dual clutch fingers can be adjusted for wear without disassembly or removal from tractor. Proceed as follows.

Disconnect clutch operating rod (R—Fig. 89A) from clutch pedal arm and refer to Fig. 89B. Remove access cover from clutch housing and back off each pair of nuts (2) until operating lever (3) contacts pto clutch plate pin (4) of all three operating levers. Tighten each of the three inner nuts until fingers (3) are just ready to pull away from release pins (4). Tighten each of the inner nuts (2) an additional 2½ turns, then tighten lock nuts. Reconnect operating linkage and

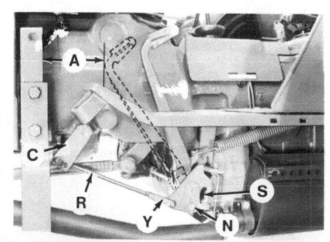

Fig. 89—On models equipped reversing transmission, clutch pedal free travel (A) is adjusted by turning yoke (Y) on rod (R) after loosening locknut and detaching one end of rod. Refer to text.

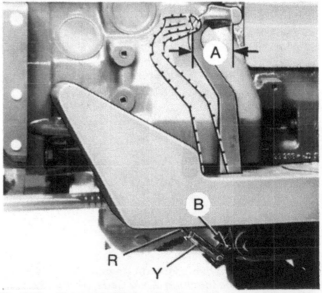

Fig. 89A—External clutch adjustment is accomplished by threading yoke (Y) onto or out of rod (R). Refer to text.

adjust pedal free travel (A—Fig. 89A) as described in paragraph 97A.

Hydraulic Linkage

98. PEDAL ADJUSTMENT. Remove dash cover attaching screws and lift off left cover. Refer to Fig. 91 and measure distance (B—Fig. 90) between floor plate and pedal pivot point, then measure distance (C) between floor plate and return spring bracket pivot point. Distance (A) is difference between measured distance (B) and measured distance (C). Distance (A) should be 24-26 mm (0.945-1.023 inches) for all 2350 models, all 2550 models, 2355 models before SN 620 388L and 2555 models before SN 620 388L. Distance (A) should be 20-22 mm (0.787-0.866 inch) for 2355 models after SN 620 387L and 2555 models after SN 620 387L. If adjustment is required, loosen lock nut and turn stop screw (2), which will change measurement (C). Lock adjustment by tightening locknut when adjustment is correct.

Move boot (7—Fig. 93 or Fig. 93A) to expose flats on operating rod (8). Loosen lock nut, turn rod in or out of clevis until piston (3) is touching washer (5) and rod (8) is still contacting piston. Tighten rod locknut against clevis. Fully depress clutch pedal and measure the distance operating rod (8) travels. Master cylinder operating rod travel should be 26-28 mm (1.024-1.102 inches) for 2350 and 2550 models; 29-31 mm (1.142-1.220 inches) for 2355 and 2555 models so equipped. Stop screw (S—Fig. 90A) can be adjusted to limit master cylinder travel.

With master cylinder and pedal adjusted, external slave cylinder travel can be checked on 2350 and 2550 models. Fully depress clutch pedal and measure the distance operating rod (12—Fig. 94) travels. This distance should be at least 8.5 mm ($\frac{5}{16}$ inch) and not more than 12 mm ($\frac{15}{32}$ inch). If distance is less than 8.5 mm ($\frac{5}{16}$ inch), bleed system as in paragraph 98A and recheck.

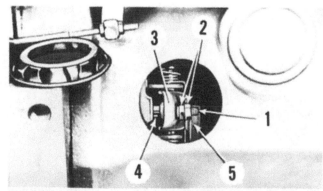

Fig. 89B—Wear adjustment of dual clutch levers can be made through opening in clutch housing after removing access cover.

1. Operating bolt
2. Jam nuts
3. Operating lever
4. Clutch plate pin

BLEEDING HYDRAULIC CLUTCH SYSTEM

All Models So Equipped

98A. To bleed air from system, first fill reservoir with DOT 4 brake fluid. DO NOT use any type of transmission or hydraulic oil. Remove cap from bleed valve (19—Fig. 91 or Fig. 91A) and attach a hose to valve. Insert hose end into a container partially filled with brake fluid. Open bleed valve ½ turn, fully

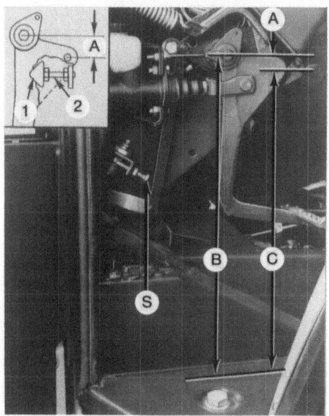

Fig. 90—View showing pedal adjustment typical of models with hydraulic operated clutch. Inset shows measurement (A), pedal stop (1) and stop bolt (2).

Fig. 90A—Stop screw (S) limits the operating range of master cylinder piston. Screw is also shown in Fig. 90.

depress clutch pedal, close bleed valve, then release pedal. Repeat this procedure until only air-free fluid flows from bleed valve as indicated by flow that is clean and free from bubbles.

If tractor is equipped with Hi-Lo shift unit, remove floor mat and floor panel. Bleed Hi-Lo locking device on transmission shift cover in similar manner.

> NOTE: When bleeding system, make sure reservoir does not run dry.

After bleeding system, make sure fluid reservoir is filled.

CLUTCH OPERATING CYLINDERS

2350 and 2550 Models So Equipped

100. R&R AND OVERHAUL. Remove dash cover attaching screws and lift off left cover. Attach a hose to bleed screw on slave cylinder and insert hose end in a suitable container. Open bleed screw one turn

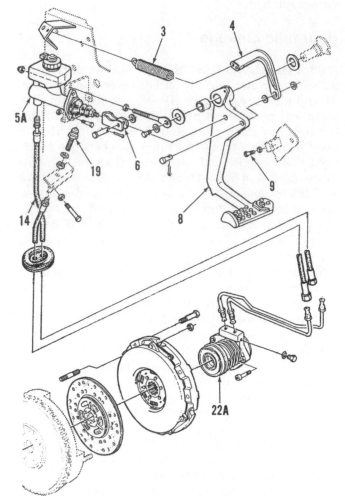

Fig. 91A— Exploded view of hydraulic clutch linkage and related parts used on 2355 and 2555 models with Sound Gard Body. Refer to Fig. 91 for legend except the following:

    5A. Master cylinder & reservoir
    22A. Throwout bearing & slave cylinder

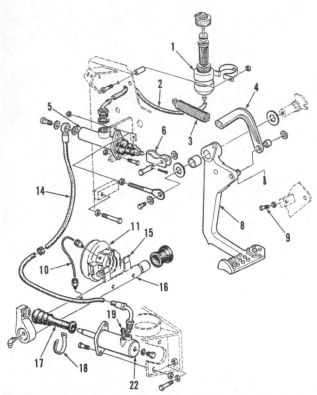

Fig. 91—View showing hydraulic clutch linkage and related parts used on 2350 and 2550 models with Sound Gard Body.

1. Reservoir	
2. Hose	11. Release bearing
3. Pedal return spring	14. Pressure hose
4. Return bracket	15. Fork
5. Master cylinder	16. Clutch release shaft
6. Clevis	17. Rubber boot
8. Clutch pedal	18. Tie band
9. Stop bolt	19. Bleeder valve
10. Lube line	22. Slave cylinder

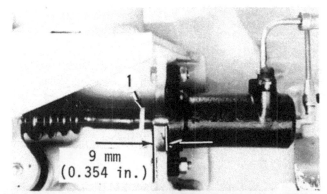

Fig. 92—Slave cylinder is installed at rear of mounting bracket on early 2350 and 2550 tractors with early style clutch release shaft and release bearing. With new clutch disc installed, secure boot to rod with tie band (1), 9 mm (0.354 inch) between end of boot and front face of cylinder.

and drain system by operating clutch pedal. Disconnect lines from cylinders, then unbolt and remove cylinders.

Procedure for disassembly of cylinders is obvious after examination of units and reference to appropriate Fig. 93 or Fig. 94. Repair kits are available for overhauling cylinders.

Reassemble and reinstall cylinders, check adjustments as in paragraph 98, then fill and bleed system as outlined in paragraph 98A.

NOTE: Refer to Figs. 92 and 92A for installation information on slave cylinders on 2350 and 2550 models.

2355 and 2555 Models So Equipped

100A. R&R AND OVERHAUL. To remove the clutch master cylinder, remove dash cover attaching screws and lift off left cover. Attach a hose to bleed screw (19—Fig. 91A) and insert hose end in a suitable container. Open bleed screw one turn and drain system by operating clutch pedal. Disconnect line from master cylinder, then unbolt and remove cylinder.

Procedure for disassembly of master cylinder is obvious after examination of units and reference to Fig. 93A. **DO NOT lubricate master cylinder with anything except DOT 4 brake fluid.** Repair kit is available. Reassemble and reinstall cylinder, check adjustments as in paragraph 98, then fill and bleed system as outlined in paragraph 98A.

To remove the slave cylinder, it is necessary to split the tractor as outlined in paragraph 101. Disconnect lines, remove screws (45—Fig. 94A), then remove the assembly. To disassemble, remove retaining ring (25) and release bearing (26). Remove snap ring (27), then slide piston (32) from guide sleeve (44). **DO NOT lubricate any of the clutch slave cylinder with anything except DOT 4 brake fluid.** Seals will be

damaged by contact with oil or grease. Repair kit is available. Reassemble and reinstall cylinder, rejoin tractor halves, then check adjustments as in paragraph 98. Fill and bleed system as outlined in paragraph 98A.

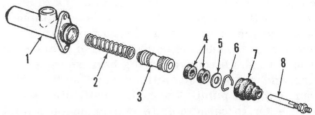

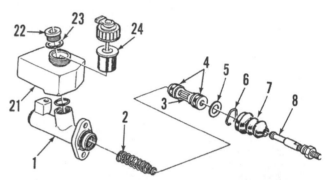

Fig. 93—Exploded view of hydraulic clutch master cylinder used on 2350 and 2550 models with Sound Gard Body. Refer to Fig. 91 for reservoir (1) and other related parts.

1. Housing	5. Washer
2. Spring	6. Snap ring
3. Piston	7. Rubber boot
4. Seal rings	8. Operating rod

Fig. 93A—Exploded view of hydraulic clutch master cylinder used on 2355 and 2555 models with Sound Gard Body.

1. Housing	7. Rubber boot
2. Spring	8. Operating rod
3. Piston	21. Reservoir
4. Seal rings	22. Screw
5. Washer	23. Washer
6. Snap ring	24. Strainer

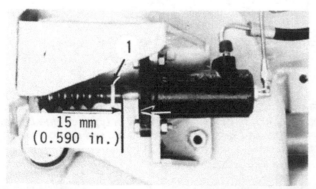

Fig. 92A—Slave cylinder is installed in front of mounting bracket of late 2350 and 2550 tractors and earlier tractors if the late style clutch release shaft and release bearing has been installed. With new clutch disc installed, secure boot to rod with tie band (1), 15 mm (0.590 inch) between end of boot and front face of cylinder.

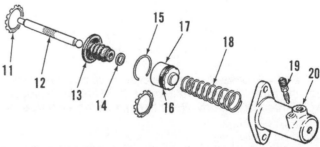

Fig. 94—Exploded view of hydraulic clutch slave cylinder used on 2350 and 2550 models with Sound Gard Body.

11. Retainer	16. Seal ring
12. Operating rod	17. Piston
13. Rubber boot	18. Spring
14. Retaining ring	19. Bleed valve
15. Snap ring	20. Housing

TRACTOR SPLIT

All Models

101. To split engine from clutch housing, drain cooling system, disconnect batteries, then remove batteries. Remove exhaust pipe, radiator side plates and grille screens. On models with Sound Gard Body, remove battery boxes and on models so equipped, disconnect air conditioning lines at unions under body. On all models, disconnect wires to lights, then remove hood. Disconnect wiring harness from dash panel and engine speed control rod and shut-off cable from injection pump. On models with front wheel drive, refer to paragraph 18 and remove the front drive shaft. Disconnect heater hoses, hydraulic lift system lines and power steering lines that would interfere with separation of tractor between clutch

housing and engine. Detach drag link from steering bellcrank from models so equipped. On models without Sound Gard Body, remove cap screws that attach dash panel at flywheel housing. On models with an auxiliary fuel tank, drain fuel from main tank, disconnect interfering hoses and remove the auxiliary fuel tank. On all models so equipped, remove the two lower screws that attach the flywheel housing to the oil pan. On all models, attach splitting stands and support both the front and rear sections of tractor in such a way to safely allow separating between clutch housing and flywheel housing. Remove platform mat and center floor panel from models with Sound Gard Body, then remove upper right cap screw through this opening. On all models, remove the cap screws and hex nuts attaching clutch housing to the flywheel housing, then separate between the two housings. **It may be necessary to use special wrench**

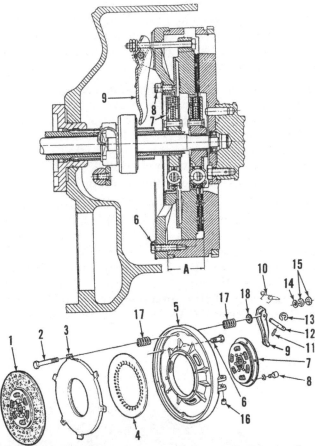

Fig. 94A—Exploded view of hydraulic clutch slave cylinder and throwout bearing used on 2355 and 2555 models with Sound Gard Body. Refer to Fig. 91A for related parts.

25. Retaining ring	
26. Throwout bearing	
27. Snap ring	36. Seal ring
28. Wiper seal	37. Outer protective cover
29. "O" ring	38. Inner protective cover
30. Guide ring	39. Spring
31. Guide ring	40. Seal ring
32. Piston	41. Housing
33. Back-up washer	42. "O" ring
34. "V" packing	43. Snap ring
35. Seal ring	44. Guide sleeve
	45. Allen screws

Fig. 96—Exploded view and cross section of single stage, 280 mm diameter, diaphragm type, engine clutch assembly and relative components used on some early models.

1. Clutch friction disk	
2. Operating bolt	10. Loop springs
3. Pressure plate	11. Spring pins
4. Diaphragm spring	12. Pivot pins
5. Clutch cover	13. Snap rings
6. Cap screws	14. Washers
7. Torsion damper	15. Nuts
8. Allen screws	16. Bushings
9. Operating levers	17. Springs
	18. Washers

(KJD10129 or equivalent) to remove the upper hex nut from the right side on tractors with Sound Gard Body.

When rejoining tractor it may be necessary to rotate engine crankshaft (using special tool JDE-83 or equivalent) to facilitate entry of input shafts into clutch discs. Be sure flywheel housing and clutch housing are butted together before tightening retaining cap screws and hex nuts. Tighten nuts and cap screws to a torque of 230 N·m (170 ft.-lbs.).

R&R AND OVERHAUL CLUTCH

Single Stage Diaphragm Spring Clutch

102. Tractors without continuous pto or direction reverser are equipped with single dry disc diaphragm spring clutch assembly. The 280 mm (11 inches) diameter F&S clutch is shown in Fig. 96; the 320 mm (12½ inches) diameter F&S clutch is shown in Fig. 97 and the 320 mm (12½ inches) diameter LUK clutch is shown in Fig. 98. Differences will be noticed and it is important to correctly identify the assembly for servicing and for ordering parts.

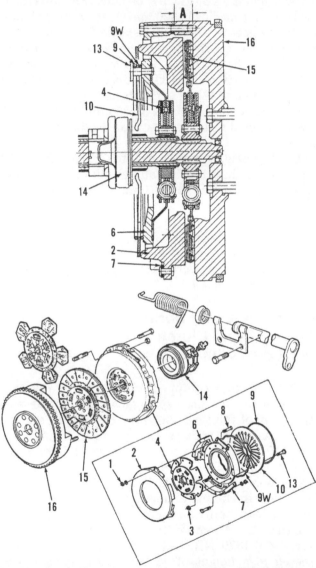

Fig. 97—Exploded view and cross section of F&S 320 mm diameter, single stage, diaphragm type engine clutch assembly. Distance "A" is depth of flywheel friction surface from clutch mounting surface.

1. Nuts
2. Pressure plate
3. Nuts
4. Torsion damper
5. Cap screw
6. Clutch cover
7. Plate springs
8. Cap screws
9. Wire ring
9W. Waved ring
10. Diaphragm spring
13. Special screws
14. Throwout bearing
15. Friction disk
16. Flywheel

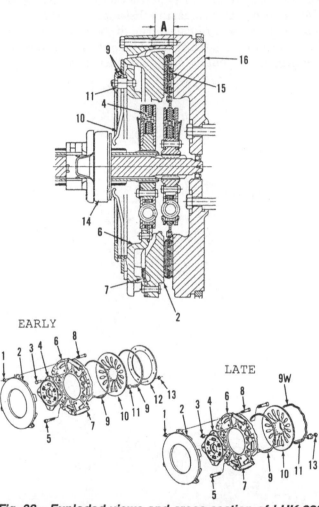

Fig. 98—Exploded views and cross section of LUK 320 mm diameter, single stage, diaphragm type engine clutch assembly. Some differences may be noted between early style and later style.

1. Nuts
2. Pressure plate
3. Rivets (early)
4. Torsion damper
5. Screws
6. Clutch cover
7. Plate springs
8. Cap screws
9. Wire ring
9W. Waved ring
10. Diaphragm spring
11. Spacer sleeves
12. Retaining ring
13. Nuts
14. Throwout bearing
15. Friction disk
16. Flywheel

To remove clutch assembly from all models, split tractor as outlined in paragraph 101, remove hex nuts and cap screws, then lift off clutch assembly.

The original F&S 280 mm (11 inches) diameter clutches (Fig. 96) used on some early models may be replaced with later 320 mm (12½ inches) diameter clutch assembly, but flywheel must also be changed. If early (280 mm) flywheel is surfaced, distance from the clutch friction surface of flywheel to clutch cover attaching surface (A—Fig. 96) should be 55.87-56.13 mm (2.20-2.21 inches). Tighten Allen screws (8) that attach torsion damper (7) to 20-25 N·m (14-18 ft.-lbs.) torque. To adjust lever height, attach magnetic spacer disks (JDE-143 or equivalent) to friction surface spaced equally around flywheel or install new clutch friction disk, then bolt clutch assembly to flywheel. Use a depth gage to measure distance from throw out bearing surface of release lever to the rear of torsion damper splines. Correct lever height is different depending upon length of the release lever. Release

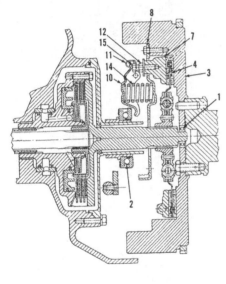

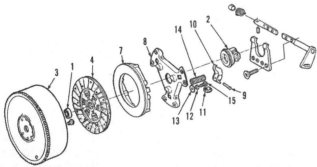

Fig. 99—Exploded view of single stage engine clutch assembly and relative components used on models with direction reverser.

1. Pilot bearing	9. Pivot pins
2. Throwout bearing	10. Release levers
3. Flywheel	11. Return clips
4. Clutch disc	12. Lock nuts
6. Cap screws	13. Washers
7. Pressure plate	14. Springs
8. Clutch cover	15. Adjusting screws

lever height should be 45.5-46.0 mm (1.791-1.811 inches) if the distance between lever pivot hole and center end of lever is 107.5 or 110.5 mm (4.23 or 4.35 inches). Release lever height should be 36.5-37 mm (1.437-1.456 inches) if the distance between lever pivot hole and center end of lever is 103 mm (1.437 inches). Change release lever height if necessary, then tighten lock nuts to 25-30 N·m (18-23 ft.-lbs.) torque. Recheck height after lock nuts are tightened. Clutch release lever height should still be within suggested limits and lever height should not be more than 0.4 mm (0.015 inch) different than other levers.

The original F&S (Fig. 97) or LUK (Fig. 98) clutch used on some models is 320 mm (12½ inches) diameter. If flywheel is surfaced, distance from the clutch friction surface of flywheel to clutch cover attaching surface (A—Fig. 97 or Fig. 98) should be 21.35-21.85 mm (0.841-0.860 inches). On LUK models, thickness of pressure plate (2) from diaphragm spring contact surface to friction surface is 53.85 mm (2.120 inches) new and should not be less than 52.85 mm (2.081 inches).

On all models, new engine clutch friction disc is 10 mm (0.390 inch) thick. Install new disc if total thickness at facing is less than 7 mm (0.260 inch). When installing, make sure side of disc marked FLYWHEEL SIDE (side with star-shaped plate around hub) is toward flywheel. Use a centering tool when installing. Remove centering tool after clutch retaining hex nuts and cap screws are tightened to a torque of 50 N·m (35 ft.-lbs.).

Single Stage Coil Spring (Auburn Type) Clutch All Models With Reverser

102A. Tractors equipped with direction reverser have a single, dry disc, spring-loaded clutch assembly typical of the type shown in Fig. 99. To remove clutch assembly, split tractor as outlined in paragraph 101, remove the six cap screws and hex nuts, then lift off clutch assembly.

New engine clutch disc is 10.7-11.3 mm (0.421-0.445 inch) thick. Install new disc if total thickness at facing is less than 7.7 mm (0.303 inch). Free length of springs is 64 mm (2.52 inches) and each should exert 1240-1370 N (279-308 lbs.) force when compressed to a height of 46 mm (1.81 inches). The friction surface of flywheel should be 37.5-37.7 mm (1.475-1.485 inch) from clutch mounting surface.

When installing, make sure long hub of clutch disc is toward rear (reverser clutch). Use a centering tool when assembling. Remove centering tool after clutch retaining hex nuts and cap screws are tightened to 50 N·m (35 ft.-lbs.) torque. The clutch release lever height should be adjusted only with a new clutch disc and special gage (JD-7 or equivalent) when clutch is assembled to the flywheel that is to be used with the clutch assembly. Measure distance from contact surface of release levers to clutch mounting surface of flywheel. Loosen locknut (12—Fig. 99A) and turn adjusting screw (15) until release lever height is correct, then tighten locknut. Recheck after adjusting to be sure all levers are the same height within recommended range. Height of the three levers should be the same after lock nuts are tightened.

The clutch is a balanced assembly. All parts should be marked to indicate original position before disassembling. To disassemble, compress clutch springs, loosen locknuts (12—Fig. 99), then remove screws (15). Release pressure from clutch springs slowly, then disassemble remaining parts as necessary.

Dual Stage Diaphragm Spring Clutch Continuous Running Pto

102B. The dual stage diaphragm clutch combines the functions of two clutches into one assembly and provides for continuous running pto. To remove clutch assembly, split tractor as outlined in paragraph 101, remove cap screws (1—Fig. 100), then lift clutch assembly from flywheel.

The clutch is a balanced assembly and all parts should be marked to indicate original position before disassembling. Remove locknuts (14J—Fig. 100), loosen adjusting nuts (14) until they have relieved spring pressure, then remove nuts (14) and screws (4). Remove snap rings (15), unhook springs (12), drive spring pins (17) from pivot pins (16) and levers (11), then push out the pivot pins (16). Inspect all parts for cracks, wear, distortion and scoring. New diaphragm spring (6) should be installed if old spring is in anyway questionable. Thickness of new engine friction disk (3) is 9-10 mm (0.366-0.390 inch) and new disk should be installed if less than 6 mm (0.238 inch). Thickness of new pto friction disk (8) is 7.5-8.5 mm (0.295-0.335 inch) and new disk should be installed if less than 4.7 mm (0.185 inch). If flywheel is surfaced, distance from the clutch friction surface of flywheel to clutch cover attaching surface (A—Fig. 100) should be 55.87-56.13 mm (2.200-2.210 inches).

When reassembling, bushing surfaces of operating bolts (4), bushing (10) and operating pins (18) should be coated with Moly High-Temperature EP grease. Also coat the rubbing surfaces on sides of pressure plates (5 and 7) with the high-temperature grease.

To adjust lever height, attach magnetic spacer disks (JDE-143 or equivalent) to friction surface spaced equally around flywheel or install new clutch friction disk, then bolt clutch assembly to flywheel. Use a depth gage to measure distance from throw out bearing surface of release lever to the hub of the rear (pto) clutch disk. Correct lever height is 37.0-37.6 mm (1.466-1.474 inches). Turn nuts (14) to change release lever setting, if necessary, then tighten lock nuts (14J)

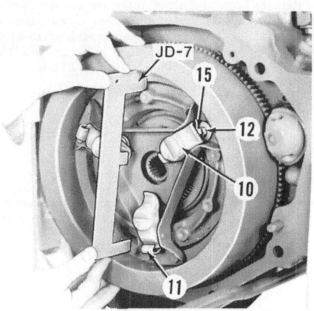

Fig. 99A—Measure release lever height as shown using special gage and new clutch friction disk as described in text.

10. Release lever	12. Lock nut
11. Return clip	15. Adjusting screw

to 25-30 N·m (18-23 ft.-lbs.) torque. Recheck height after lock nuts are tightened. Clutch release lever height should still be within suggested limits and lever height should not be more than 0.4 mm (0.015 inch) different than other levers. Use a centering tool (JDE-52-A or equivalent) when installing clutch and friction disks. Side of engine clutch disk (3) marked "FLYWHEEL SIDE" should have long hub and should be toward flywheel. Remove centering tool after clutch retaining cap screws (1—Fig. 100) are tightened to 50 N·m (35 ft.-lbs.) torque.

CLUTCH RELEASE BEARING AND SHAFT

All Models Except 2355 and 2555 With Sound Gard Body

104. REMOVE AND REINSTALL. Separate engine from clutch housing as outlined in paragraph 101. On models so equipped, remove lube tube and fittings. Slide release bearing from carrier. Unbolt fork from release shaft. On models so equipped, disconnect mechanical linkage rod from shaft arm. On models with hydraulic clutch linkage, remove rubber boot from clutch shaft arm. Slide release shaft out left side of clutch housing on all models.

Inspect and renew release shaft bushings as necessary. Drive plastic-coated bushings in until flush with clutch housing. **Do not lubricate plastic coated bushings with graphite grease.**

Reinstall components by reversing removal procedure. On models with mechanical clutch linkage, hook short leg of return spring against clutch fork and long leg against clutch housing. Apply Loctite 271 to fork retaining cap screws and tighten to a torque of 50 N·m (35 ft.-lbs.).

Models 2355 and 2555
With Sound Gard Body

104A. REMOVE AND REINSTALL. Separate engine from clutch housing as outlined in paragraph 101. Refer to paragraph 100A for service to the clutch

throwout bearing and hydraulic operating system used on these models.

CLUTCH HOUSING

All Models Without Direction Reverser

105. Clutch housing normally will not need complete removal for servicing. Clutch control linkage can be serviced when clutch housing is separated from engine. Clutch shaft and pto shaft along with their bearings and oil seals and the transmission oil pump can be serviced after clutch housing is separated from transmission case.

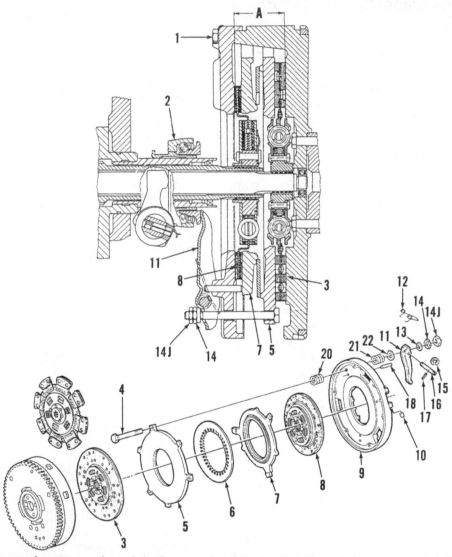

Fig. 100—Exploded view and cross section of dual stage clutch typical of all models so equipped. Some models may not be equipped with parts (20, 21 and 22) as shown in the cross section. Parts (20, 21 and 22) are shown installed in Fig. 100A.

1. Cap screws	8. Pto friction disk	13. Spacer washer	17. Spring pins
3. Engine clutch friction disk	9. Clutch cover	14. Adjusting nuts	18. Release pins
4. Operating bolts	10. Bushings	14J. Lock nuts	20. Springs (44 mm)
5. Pressure plate (engine clutch)	11. Release levers	15. Snap rings	21. Springs (41.5 mm)
6. Diaphragm spring	12. Loop springs	16. Pivot pins	22. Seat disks
7. Pressure plate (pto clutch)			

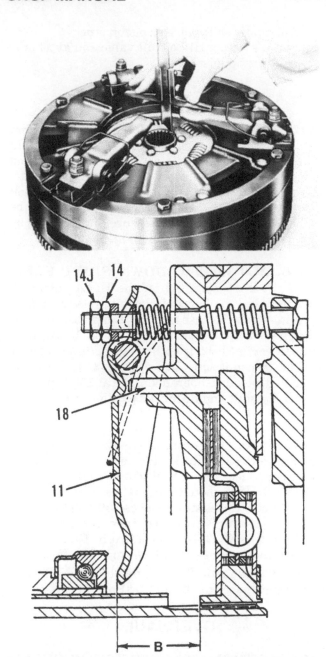

Fig. 100A—Distance "B" in the cross section is measured between release levers and rear edge of pto clutch friction disk hub using a depth gage. Change distance by turning adjusting nut (14) and lock position by tightening lock nut (14J).

Refer to appropriate paragraphs for service information of components contained in the housing.

HI-LO SHIFT UNIT

OPERATION

106. The Hi-Lo shift unit is a hydraulically shifted gear reduction unit that can be shifted under load in any transmission gear while on the go without declutching. Shifting the Hi-Lo unit from Hi to Lo slows tractor ground speed approximately 20 percent and at the same time increases torque to tractor rear axle. Shifting back to Hi, resumes normal ground travel speed.

The Hi-Lo unit is located directly in front of the transmission in the rear of clutch housing.

The transmission oil pump supplies lubricating and pressurized oil to the Hi-Lo unit as well as the front-wheel drive clutch, the pto clutch and is also used as a charge pump for the main hydraulic system pump. Oil from the transmission pump passes through the transmission oil filter, then to the pressure regulating valve, automatic shift valve and Hi-Lo control valve in the transmission shift cover. The transmission oil pressure indicator light should come on when pressure drops to less than about 930 kPa (135 psi) and should go off when pressure increases to about 965 kPa (140 psi). The Hi-Lo shift valve is designed so that if there is a drop in system oil pressure below 750-850 kPa (109-116 psi), there will be an automatic shift to Lo to prevent slippage and possible damage to the Hi Clutch discs (1 and 2—Fig. 103).

When Hi-Lo control lever is shifted to Hi, pressure oil flows behind Hi Clutch piston (24) and Lo Brake piston (35). Piston (24) overcomes the pressure of Belleville springs (25) and applies Hi Clutch (1 and 2). Brake piston (35) overcomes the pressure of Belleville springs (16) freeing up Lo Brake discs (38 and 39). This allows the drive shaft, sun gear (5), planet carrier, output sun gear (4) and transmission input shaft (42) to rotate as a single unit at engine speed for direct drive.

When control lever is shifted to Lo, pressure oil is removed from behind pistons (24 and 35). Belleville springs (25) force piston (24) back to release Hi Clutch discs (1 and 2). Belleville springs (16) force piston (35) back and applies pressure to Lo Brake discs (37 and 39). Power now flows through unit via drive (engine clutch) shaft, input sun gear (5), planet gears (13), output sun gear (4) and transmission input shaft (42).

NOTE: When operating in Hi, if operating pressure drops below 750-850 kPa (109-116 psi) or engine is shut off, Hi-Lo unit will automatically shift to Lo speed position.

TROUBLE-SHOOTING

107. Some problems that may occur during operation of the Hi-Lo unit and their possible causes are as follows:

1. Hi-Lo control lever jumping out of engagement. Could be caused by:
   a. Transmission oil filter clogged.
   b. Transmission oil pressure too low.
   c. Pressure regulating spring broken.

2. Transmission oil pressure indicator light glows. Could be caused by:
   a. Faulty transmission oil pump.
   b. Low transmission oil level.
   c. Pressure regulating valve set too low.
   d. Transmission oil filter clogged.

3. Hi-Lo speed of shift too slow. Could be caused by:
   a. Transmission oil pressure too low.
   b. Worn disc pack.
   c. Mechanical failure in Hi-Lo unit.

4. Control lever shifts below specified pressure or does not shift to Hi. Could be caused by:
   a. Faulty gage used for checking pressure.
   b. Automatic shift valve or piston stuck.
   c. Check valve at front of down shift valve improperly installed.

5. Control lever shifts above specified pressure. Could be caused by:
   a. Faulty gage used for checking pressure.
   b. Incorrect assembly of automatic shift valve.
   c. Broken or weak spring at automatic shift valve or piston.

d. Automatic shift valve or piston stuck.
e. Pin connecting Hi-Lo shift valve and automatic shift piston missing.

PRESSURE TEST

107A. SYSTEM PRESSURE. To check system operating pressure, remove plug (42—Fig. 102) and install a 2070 kPa (300 psi) test gage. Place range and gear shift levers in neutral and apply hand brake. Operate engine at 1500 rpm and warm transmission oil to 65° C (150° F). Operating pressure should be 1050 kPa (150 psi). Adjust pressure, if necessary, by adding or removing shims (64—Fig. 104).

107B. AUTOMATIC DOWN SHIFT PRESSURE. To check operation of the Hi-Lo automatic down shift and the transmission oil pressure indicator light, attach the hose to a flow and pressure meter to the ported filter cover port (B—Fig. 102A), then direct the outlet hose from flow and pressure meter into the transmission filler opening (A). Close the meter control valve, operate engine at 1500 rpm and move Hi-Lo shift handle to "Hi" position. Slowly open the meter control valve while observing the low pressure indicator light and pressure at which the automatic shift to "Lo" occurs. The transmission oil pressure indicator light should come on when pressure drops to less than about 930 kPa (135 psi). The Hi-Lo shift valve should automatically shift to "Lo" when system oil pressure reaches 750-850 kPa (109-116 psi). If pressures are incorrect, refer to TROUBLE-SHOOTING paragraph 107. System pressure will increase after lever moves to "Lo."

The low pressure indicator light should go off when pressure increases to about 965 kPa (140 psi).

OVERHAUL

108. CONTROL AND PRESSURE REGULATING VALVES. Control valve, automatic down shift valve and pressure regulating valve are located in transmission shifter cover as shown in Fig. 104. Regulating valve can be adjusted without removing shifter cover, but valve unit overhaul can be accomplished only after cover is removed.

On models without Sound Gard Body, unbolt and remove transmission shield by working it up over shift lever boots. Disconnect Hi-Lo shift lever linkage and rear wiring harness. Unbolt and remove shift cover, valves and shift levers.

On models with Sound Gard Body, remove floor mat and platform center section. Disconnect Hi-Lo shift linkage and rear wiring harness. Disconnect line from shift lock elbow, then unbolt and remove cover and valve assembly. Unbolt and remove shift locking device, if so equipped.

Fig. 102—Hi-Lo system pressure is checked at port where plug (42) is installed. Pressure regulating valve plug is shown at (1). Refer to text.

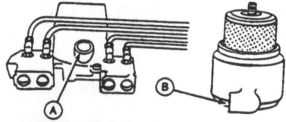

Fig. 102A—Automatic down shift of the Hi-Lo control valve can be checked with flow-pressure meter as described in text.

On all models, unbolt and remove housing (32). Remove detent assembly (37 through 40), then remove control valve (48).

CAUTION: Plugs (56 and 62) are under spring force, use caution when removing them.

Remove plug (56) and automatic down shift valve (55). Remove plug (62) and regulating valve (68).

Check springs and valves against the following specifications:

Regulating valve outer spring (65)—
Free length . . . . .129 mm
(5.079 in.)
Test length . . . . .64 mm
(2.52 in.)
Test load .95-120 N
(22-26 lbs.)

Regulating valve inner spring (66)—
Free length . . . . .182 mm
(7.165 in.)
Test length . . . . .109 mm
(4.291 in.)
Test load .85-105 N
(19-23 lbs.)

Control valve spring (53)—
Free length . . . . 71 mm
(2.795 in.)
Test length . . . . 43 mm
(1.701 in.)
Test load .220-270 N
(49-60 lbs.)

Control valve detent spring (39)—
Free length . . . . 49.5 mm
(1.949 in.)
Test length . . . . 36.5 mm
(1.429 in.)
Test load .50-60 N
(11-13 lbs.)

Automatic down shift spring (58)—
Free length . . . . 71 mm
(2.795 in.)
Test length . . . . 43.5 mm
(1.713 in.)
Test load .55-65 N
(12-15 lbs.)

Control valve (54)—
Spool OD . . . . 12.65-12.67 mm
(0.498-0.499 in.)
Bore ID . . . . 12.73-12.74 mm
(0.5013-0.5014 in.)

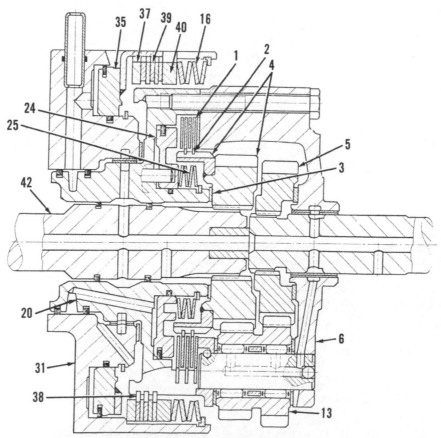

Fig. 103—Sectional view of typical Hi-Lo unit. Refer to Fig. 106 for legend.

Control valve piston (51)—
Piston OD . . . . 20.44-20.46 mm
(0.804-0.805 in.)
Bore ID . . . . 20.50-20.53 mm
(0.807-0.808 in.)

Regulating valve (67)—
Plunger OD . . . . 15.95-15.97 mm
(0.628-0.629 in.)
Bore ID . . . . 16.00-16.01 mm
(0.629-0.630 in.)

Automatic down shift valve (59)—
Spool OD . . . . 8.96-8.97 mm
(0.352-0.353 in.)
Bore ID . . . . 9.00-9.02 mm
(0.354-0.355 in.)

All springs are available separately, however, if valves or bores are excessively worn or damaged, shift cover (41) with valves must be renewed. Inspect locking device, if so equipped, for damage or leakage and renew parts as necessary. The locking device prevents

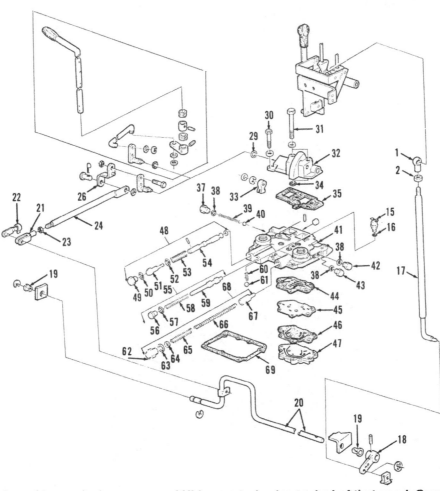

Fig. 104—Exploded view of transmission cover and Hi-Lo control valve typical of that used. Some items are only used on remote shifting for models equipped with Sound Gard Body.

1. Ball socket	28. Cap screw		57. "O" ring
2. Locknut	29. "O" ring	43. Plug	58. Spring
15. Oil pressure switch	30. Cap screw	44. Gasket	59. Valve spool
16. "O" ring	31. Cap screw	45. Plate	60. Spring
17. Connecting rod	32. Housing	46. Gasket	61. Ball
18. Levers	33. Arm	47. Manifold cover	62. Plug
19. Bushings	34. "O" ring	48. Hi-Lo control valve	63. "O" ring
20. Shaft	35. Gasket	49. Plug	64. Shim
21. Yokes	36. Detent assy.	50. "O" ring	65. Spring (outer)
22. Retainers	37. Plug	51. Piston	66. Spring (inner)
23. Locknuts	38. "O" rings	52. "O" ring	67. Valve plunger
24. Rod	39. Spring	53. Spring	68. Regulating valve
25. Nut	40. Ball	54. Spool	69. Gasket
26. Lever	41. Cover	55. Automatic shift valve	70. Locking device
27. Shaft	42. Plug	56. Plug	

Hi-Lo unit from down shifting when clutch pedal is depressed, on Sound Gard Body tractors, when changing transmission gears. Fluid from clutch slave cylinder forces piston and lock pin downward to lock spool (54) in position.

Renew "O" rings and gaskets and reassemble and reinstall by reversing disassembly and removal procedures.

109. HI-LO DRIVE UNIT. To remove the Hi-Lo unit, first remove the Sound Gard Body as outlined in paragraph 186, if so equipped. Remove transmission shift cover assembly as in paragraph 108, then split tractor between clutch housing and transmission as outlined in paragraph 127.

Pull transmission input shaft (1—Fig. 105) out of Hi-Lo unit. Remove oil lines (3, 4 and 7), then remove cap screws (2) and lift out Hi-Lo unit.

Set unit on a bench so that brake housing (31—Fig. 106) is facing downward. Lift planetary with Hi Clutch assembly out of brake housing and lay assem-

bly aside. Use spring compressor JDT-24A and adapter JDT-24-2 or equivalent to compress Belleville

Fig. 105—View into rear of clutch housing, showing Hi-Lo unit, pto clutch and oil lines.

1. Transmission input shaft
2. Cap screws (4)
3. Lubricating line to oil cooler
4. Inlet line from filter
5. Oil pump pressure line
6. Pto clutch
7. Pressure line to front-wheel drive clutch
8. Oil pump suction line

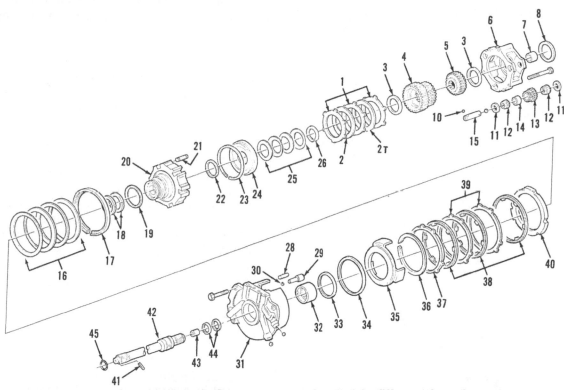

Fig. 106—Exploded view of typical Hi-Lo unit. Some parts may be slightly different than shown.

1. Hi Clutch plates (external tangs)	12. Needle bearing (6)	24. Clutch piston	35. Brake piston
2. Hi Clutch discs (internal splines)	13. Planet gear (3)	25. Hi Clutch Belleville springs	36. Snap ring
2T. Hi Clutch disc (internal splines)	14. Spacer (4)	26. Snap ring	37. Back plate
3. Thrust washers	15. Pin (3)	27. Cap screw (4)	38. Brake discs (internal tangs)
4. Sun gear & clutch hub	16. Lo Brake Belleville springs	28. Dowel pin (2)	39. Brake plates (external tangs)
5. Input sun gear	17. Snap ring	29. Priority valve	40. Brake release plate
6. Planet carrier	18. Seal rings	30. Valve ball	41. Spring pin
7. Bushing	19. Thrust washer	31. Brake housing	42. Transmission input shaft
8. Thrust washer	20. Clutch drum	32. Bushing	43. Bushing
9. Cap screw (3)	21. Dowel pin (3)	33. Seal ring	44. Seal rings
10. Lock ball (3)	22. Seal ring	34. Piston ring	45. "O" ring
11. Thrust washer (6)	23. Piston ring		

springs (16) and remove snap ring (17). Release pressure from Belleville springs, remove springs, release plate (40), brake discs (38) and plates (37 and 39). Remove snap ring (36), piston (35) with piston ring (34) and seal ring (33). Bushing (32) can now be removed if excessively worn or damaged.

Remove seal rings (18) and thrust washer (19) from clutch drum (20). Remove three cap screws (9) and place assembly on a bench with hub of clutch drum (20) facing upward. Remove clutch drum from planet carrier, then remove Hi Clutch plates and discs (1 and 2), thrust washer (3), sun gear and clutch hub (4), input sun gear (5) and thrust washer (3). Push pins (15) upward, remove lock balls (10) and remove pins. Remove planet gears (13) with needle bearings (12), spacers (14) and thrust washers (11). Use spring compressor JDT-24A or equivalent to compress Belleville springs (25), then remove snap ring (26) and springs. Use compressed air, if necessary, to remove piston (24) and piston ring (23). Remove seal ring (22) from clutch drum. Remove bushing (7) from planet carrier if excessively worn or damaged.

Clean and inspect all parts and renew any showing excessive wear or other damage. Check height of Belleville springs against the following specifications:

Brake Belleville spring (16) . . . . 5.61 mm
(0.221 in.)
Clutch Belleville spring (25) . . . . 3.23 mm
(0.127 in.)

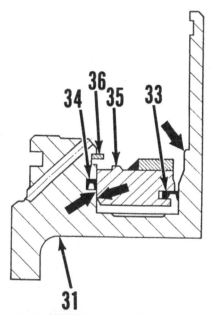

Fig. 106A—Cross section showing correct installation of seals (33 and 34). Check for burrs and sharp edges at points indicated by arrows and lubricate with multipurpose grease before assembling.

Check clutch discs and plates and brake discs and plates against the following thickness specifications:

Brake back plate (37) . . . . 4.37-4.63 mm
(0.172-0.182 in.)
Brake plate, external tangs (39) . . . . 2.30 mm
(0.091 in.)
Brake disc, internal splines (38) . . . . . 2.21-2.37 mm
(0.087-0.093 in.)
Clutch plate, external tangs (1) . . . . 1.50 mm
(0.059 in.)
Clutch disc, internal splines (2)—
Standard thickness . . . . 2.44-2.54 mm
(0.096-0.100 in.)
Special thin disc . . . . 1.55-1.60 mm
(0.061-0.063 in.)

Inside diameter of new bushing (7) is 32.11-32.16 mm (1.264-1.266 inches). Inside diameter of new bushing (32) is 85.27-85.30 mm (3.357-3.358 inches).

Use Fig. 103 and Fig. 106 as a guide and reassemble by reversing disassembly procedure, keeping the following points in mind: When installing Belleville springs (16), install with concave side toward release plate (40), then alternate concave side up, concave down and concave up toward snap ring (17). When installing clutch piston (24), make certain dowel pin in clutch drum (20) enters hole in piston. When installing Belleville springs (25), start with convex side toward piston (24), then alternate convex side up, convex down and convex up toward snap ring (26). On 2150, 2155, 2255 and 2355 models, one of the internally splined Hi clutch discs (2) is thinner than the other. The thinner disc should be installed closer to planet carrier (6). On all other models, the two discs are of the same thickness. To time planet gears, align timing marks on input sun gear (5) with timing marks on planet gears (13). Tighten planet carrier to clutch drum cap screws (9) to a torque of 30 N·m (23 ft.-lbs.).

Reinstall Hi-Lo unit in clutch housing making certain thrust washer (8) is on clutch (drive) shaft.

NOTE: The pto oil inlet line must be guided into the port in lower right front side of brake housing (31) as Hi-Lo unit is being installed.

Use Loctite 271 on cap screws (2—Fig. 105) and tighten to 50 N·m (35 ft.-lbs.) torque. Install transmission input shaft and connect oil lines. Reconnect tractor and tighten clutch housing to transmission cap screws to 160 N·m (120 ft.-lbs.) torque. Complete installation by reversing removal procedure. Reinstall Sound Gard Body, if so equipped.

DIRECTION REVERSER

OPERATION

Some tractors may be optionally equipped with a Reverser, which consists of a planetary gear set, a hydraulically actuated forward drive clutch and hydraulically actuated reverse drive brake, all operated by a remote-mounted directional lever. Pressure oil to operate the reverser unit is supplied by the transmission oil pump, which also provides lubrication for the transmission gears and serves as a supply pump for the main hydraulic system pump.

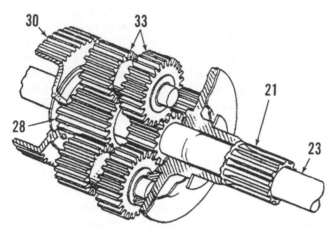

Fig. 108—Three-quarter view of direction reverser planetary unit showing component parts.

21.	Planet carrier shaft		
23.	Clutch shaft		
28.	Secondary sun gear	30.	Reverse brake hub
		33.	Planet pinions

Models So Equipped

110. Refer to Fig. 108 for a schematic view of the planetary gear unit. Planet carrier drive shaft (21) is splined into hub of forward clutch carrier. When forward clutch is engaged, clutch shaft is locked to carrier and the planetary unit turns together driving tractor in a forward direction.

The reverse brake unit is carried in transmission pump mounting plate and output sun gear (28) is splined into reverse brake hub. When reverse brake is applied, power entering at planet carrier causes the small planet pinion to walk around the fixed secondary sun gear (28) and the large planet pinion drives clutch shaft (23) at a slight overdrive in a reverse direction.

The hydraulic control unit contains a shift valve (22—Fig. 109) that directs pressurized fluid to either the forward clutch or the reverse brake piston. No neutral position is provided by the shift valve; however, included in the valve housing is a clutch control valve (25) that is moved when the clutch pedal is depressed a slight amount. This movement of the control valve will interrupt the flow of pressurized fluid to the shift valve, which will then stop forward or reverse motion of the tractor and permit transmission to be shifted normally. Design of the mechanical control linkage prevents moving the shift lever to reverse position when transmission range shifter is in "HIGH" position.

Also included in the valve housing is a spring-loaded accumulator piston (26) that smooths the engagement during direction changes. An orifice screw (30) is provided to permit adjustment of the shift rate.

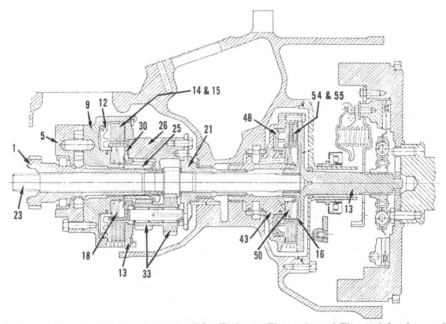

Fig. 108A—Cross section of direction reverser assembly. Refer to Fig. 112 and Fig. 114 for legend.

The accumulator remains charged when power flow is interrupted by the clutch pedal, thus permitting immediate and accurate control of clutch reengagement rate.

CHECK AND ADJUST

All Models So Equipped

111. Before making any tests, first check to make sure that transmission oil is at correct level and that transmission is at operating temperature. It is generally advisable to check and adjust the complete system as described in the following paragraphs 111A through 111E when adjustment of any part of the system is questioned.

111A. REGULATING VALVE. Remove plug (29—Fig. 109) from the bottom of cover (28) and connect a 0-2070 kPa (0-300 psi) test gage to the port. With transmission in neutral and engine running at 2100 rpm, pressure should be 1170-1240 kPa (170-180 psi). If pressure is not within range, remove plug (1) and add or remove shims (7) as required. Record the final pressure for reference during the following check described in paragraph 111B. Refer to paragraph 111E and check pedal free travel before removing test gage from this port.

111B. CLUTCH CONTROL VALVE. Make sure that regulated pressure is within range and that exact pressure is known as described in preceding paragraph 111A. Remove either plug (13 or 14—Fig. 109) from top of valve body and attach a 0-2070 kPa (0-300 psi) test gage to the port. Start engine and move reverser control lever to pressurize the circuit containing the gage. The circuit pressure should be the same as system pressure previously tested (as in paragraph 111A) and recorded. Remove gage, install plug, then remove other plug (13 or 14), attach test gage to the exposed port and check pressure of other circuit in a similar manner. Lower than system pressure indicates internal leakage of offending circuit that should be corrected by removing the leaking component and resealing or repairing.

111C. SHIFT ENGAGEMENT VALVE. Shift engagement rate when changing direction of travel can be adjusted within specific limits. Nominal shift time is ¾ to 1¼ seconds. Shift engagement rate is changed by turning the adjusting screw (30—Fig. 109) IN to slow shift rate or OUT to speed the shift rate. Initial setting is two turns out from closed (seated) position. A slow shift rate accompanied by a jerky start could indicate a broken accumulator spring or sticking accumulator piston (26).

111D. HIGH SPEED LOCK OUT. To adjust the high speed lock out, move the reverser control lever to "REVERSE," move the range shifter lever to "HIGH," then refer to Fig. 110A. Clearance (C) measured with a feeler gage should be 1.5 mm (0.060 inch).

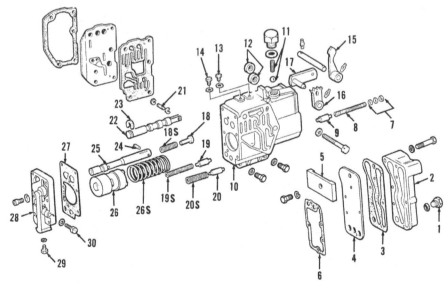

Fig. 109—Exploded view of reverser control valve showing component parts.

1. Plug	10. Body	18S. Control valve spring	24. Pin
2. Cover	11. Detent ball & spring	19. Cooler by-pass valve	25. Control valve
3. Gasket	12. Oil seals	19S. By-pass spring	26. Accumulator piston
4. Plate	13. Plug	20. Lubrication valve	26S. Accumulator piston spring
5. Block	14. Plug	20S. Lubrication valve spring	27. Gasket
6. Gasket	15. Arm	21. Pin	28. Cover
7. Shims	16. Detent arm	22. Shift valve	29. Plug
8. Spring	17. Shift shaft	23. Retaining ring	30. Adjusting screw
9. Regulating valve	18. Valve stop		

If clearance is incorrect, remove pin (C—Fig. 110B) and disconnect lower end of link rod. Turn yoke (B) as required to correct the clearance measured in Fig. 110A when yoke is reconnected and pin is reinstalled.

111E. PEDAL FREE TRAVEL. If gage is not already connected, remove plug (29—Fig. 109) from the bottom of cover (28) and connect a 0-2070 kPa (0-300 psi) test gage to the port. With transmission in neutral and engine running at 2100 rpm, pressure should be 1170-1240 kPa (170-180 psi). If pressure is not within range, refer to paragraph 111A and correct pressure. Slowly depress the clutch pedal and notice pedal position (D—Fig. 110) when pressure drops. Free travel (D) should be 19-32 mm (³⁄₄ to 1¹⁄₄ inches). If incorrect, loosen the lock screw (N) and turn screw (S) until free travel is within correct range. Tighten lock screw (N), stop engine, remove gage and install plug (29—Fig. 109).

Free travel of pedal for engine clutch is controlled by the length of rod (R—Fig. 110C) which can be changed by threading yoke (Y) onto end of rod. Correct free travel of engine clutch is correct if distance (A) is 133 mm (5¹⁄₄ inches) when forward end of rod reaches rear of slotted hole in clutch operating cross shaft and further movement of pedal will begin to move the cross shaft.

OVERHAUL

All Models So Equipped

112. REVERSER CONTROL VALVE. To remove the reverser control valve, remove the right platform, thoroughly clean the area all around the control valve and brake valve and drain the transmission case. Disconnect brake lines from rear of brake valve hous-

ing and disconnect oil cooler lines from front cover (2—Fig. 109). Remove attaching screws, then lift brake valve from tractor. Disconnect control rod yoke (B—Fig. 110B) from valve; then unbolt and remove reverser control valve.

Refer to Fig. 109 for exploded view of reverser valve. Loosen cap screws retaining rear cover (28) evenly because cover is under heavy pressure from

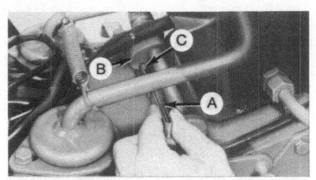

Fig. 110A—Check and adjust high speed lockout as described in text.

A. Feeler gage   B. Lockout arm   C. Clearance

Fig. 110B—Adjust link rod yoke to obtain proper clearance for high speed lockout.

A. Shift valve arm   B. Yoke   C. Pin

Fig. 110—Clutch pedal linkage showing points of adjustment and checking reverser control valve travel. Refer to text.

D. Free travel   N. Lock screw   S. Adjusting screw

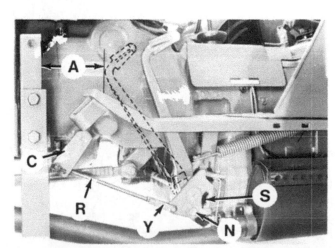

Fig. 110C—Change pedal free play for engine clutch on models with reverser by adjusting linkage as described in text.

the accumulator spring. **Cooler bypass valve (19), lubrication valve (20) and pressure regulating valve (9) are alike, but should not be interchanged after once operating in a specific bore. Springs for the valves are different** and can be identified by the following test specifications.

Pressure regulating valve spring (8)—
Free length . . . . 83.3 mm
(3.28 inches)
Tension at 65 mm . . . . 118-144.5 N
Tension at 2.56 inches . . . . 26.5-32.5 lbs.

Lubrication reduction valve spring (20S)—
Free length . . . . 52.3 mm
(2.06 inches)

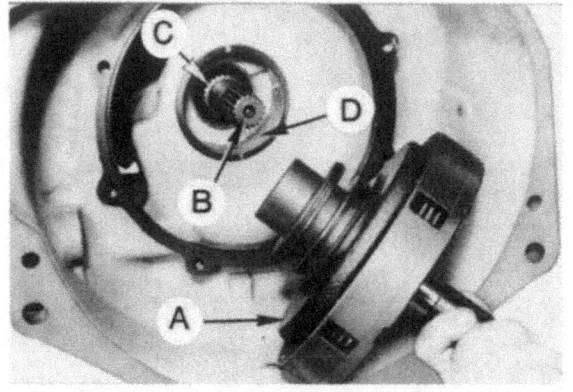

Fig. 111—Forward clutch pack can be removed as shown, after detaching engine from clutch housing.

Tension at 19.1 mm . . . . .18-22 N
Tension at 0.75 inch . . . . 4-5 lbs.

Cooler bypass valve spring (19S)—
Free length . . . . 71.4 mm
(2.81 inches)
Tension at 52.3 mm . . . . .80-98 N
Tension at 2.06 inches . . . . 18-22 lbs.

Control valve spring (18S)—
Free length . . . . 41.28 mm
(1.625 inches)
Tension at 28.58 mm . . . . .133-160 N
Tension at 1.125 inches . . . . .30-36 lbs

Detent spring (11)—
Free length . . . . 29.5 mm
(1.16 inches)
Tension at 20.6 mm . . . . 24.5-31 N
Tension at 0.81 inch . . . . .5.5-7 lbs.

Accumulator spring (26S)—
Free length . . . . 92.08 mm
(3.625 inches)
Tension at 66.68 mm . . . . .1254-1530 N
Tension at 2.625 inches . . . . 282-344 lbs.
Control valve (25) OD . . . . 17.40-17.43 mm
(0.685-0.686 inch)
Shift valve (22) OD . . . . 17.40-17.43 mm
(0.685-0.686 inch)

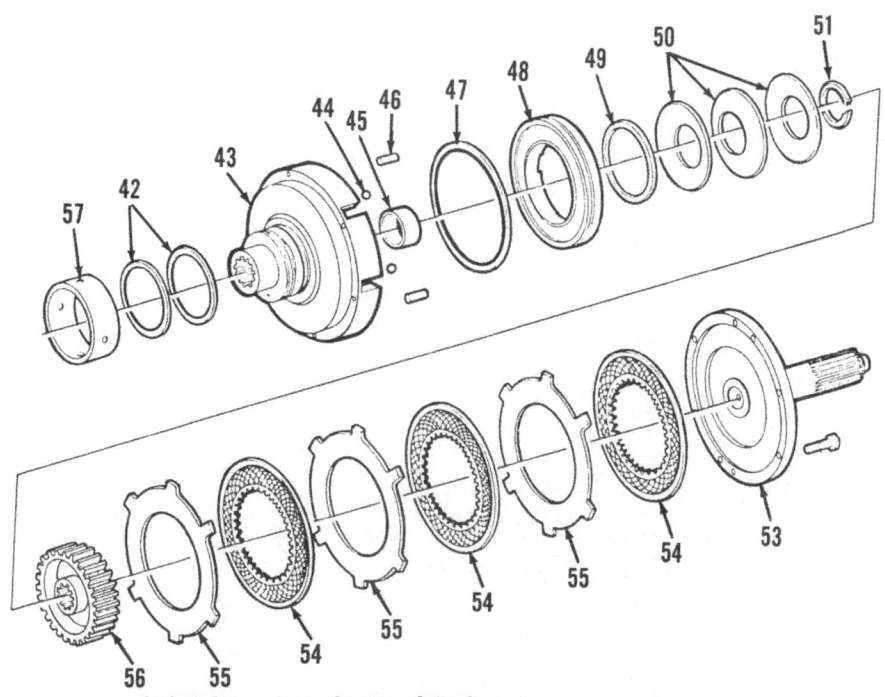

Fig. 112—Exploded view of forward clutch pack and associated parts.

42. Sealing rings	46. Dowel pin		
43. Clutch drum	47. Piston ring	50. Belleville washers	54. Clutch discs
44. Steel ball	48. Piston	51. Snap ring	55. Separator plates
45. Bushing	49. Piston ring	53. Drive shaft	56. Clutch hub

Other valves (9, 19 and 20) OD .... 12.64-12.66 mm
(0.497-0.498 inch)

Assemble by reversing disassembly procedure, making sure valves and springs are installed in correct locations. Tighten screws retaining control valve cover (28) to 15 N·m (10 ft.-lbs.); rear cover (2) to 8 N·m (6 ft.-lbs.); and valve to clutch housing to 50 N·m (35 ft.-lbs.) torque.

113. FORWARD CLUTCH. Forward clutch unit can be removed after splitting engine from clutch housing as outlined in paragraph 101. Unbolt clutch fork from cross shaft, then remove the clutch fork and throwout bearing. Disconnect rod (R—Fig. 110C), then withdraw clutch operating cross shaft from housing. Unbolt and remove the throwout bearing carrier from clutch housing, then withdraw forward clutch pack as shown in Fig. 111.

Unbolt clutch drive shaft (53—Fig. 112) from drum (43). Clutch friction discs (54), separator plates (55) and hub can now be removed. Refer to paragraph 115 for compressing Belleville washers (50) and removing snap ring (51) so that piston (48) and seals (47 and 49) can be renewed. Refer to the following specifications.

Clutch friction discs (54)—
   Thickness, new .... 2.84-3.00 mm
        (0.112-0.118 inch)
     Wear limit .... 2.79 mm
        (0.110 inch)
   Groove depth, new .... 0.25-0.51 mm
        (0.010-0.020 inch)
     Wear limit .... 0.20 mm
        (0.008 inch)

Separator plates (55)—
   Thickness, new .... 2.16-2.41 mm
        (0.085-0.095 inch)
Drum bushing (45)—
   Inside diameter, new .... 39.80-39.90 mm
        (1.567-1.579 inches)

Belleville springs (50)—
   Free height, new .... 4.65 mm
        (0.183 inch)
   Minimum limit .... 4.14 mm
        (0.163 inch)

Screws attaching drive shaft (53) to drum (43) and screws attaching throwout bearing carrier to clutch housing should all be tightened to 25 N·m (20 ft.-lbs.) torque.

114. REVERSE BRAKE AND PLANETARY UNIT. To remove the reverse brake and planetary unit, first separate the transmission from the clutch housing as outlined in paragraph 121. Transmission oil pump, reverse brake assembly and planetary unit can be removed as a unit as shown in Fig. 113; or the oil pump can be removed separately by removing pump housing cap screws. To remove either unit, it is first necessary to remove the brake valve and reverser control valve assembly as outlined in paragraph 112. Disconnect oil lines, remove retaining screws, then lift assembly out as shown in Fig. 113.

Refer to Fig. 114 for exploded view of reverse brake and planetary unit. Withdraw planet carrier (26), clutch shaft (23) and associated parts forward out of brake assembly. Withdraw clutch shaft to rear out of planet carrier.

Remove the three cap screws retaining carrier shaft (21) to carrier (26). Push each pinion shaft (35) forward slightly, remove and save locking balls (25), then remove shaft (35) and pinion (33), being careful not to lose the 44 loose needle rollers (32) in each

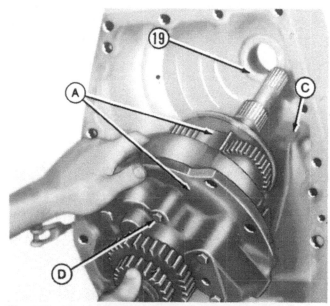

Fig. 113—Transmission oil pump (D), reverse brake and planetary unit (A) can be removed as an assembly after detaching transmission from clutch housing. Lube line is shown at (C) and thrust washer at (19).

Fig. 113A—Transmission oil pump, reverse brake and planetary unit is attached to clutch housing by screws (A).

pinion, thrust washers (31) and spacer (34). Parts (13, 14 and 15) can be removed after removing snap ring (29). Refer to paragraph 115 for compressing Belleville washers (18) and removing snap ring (17)

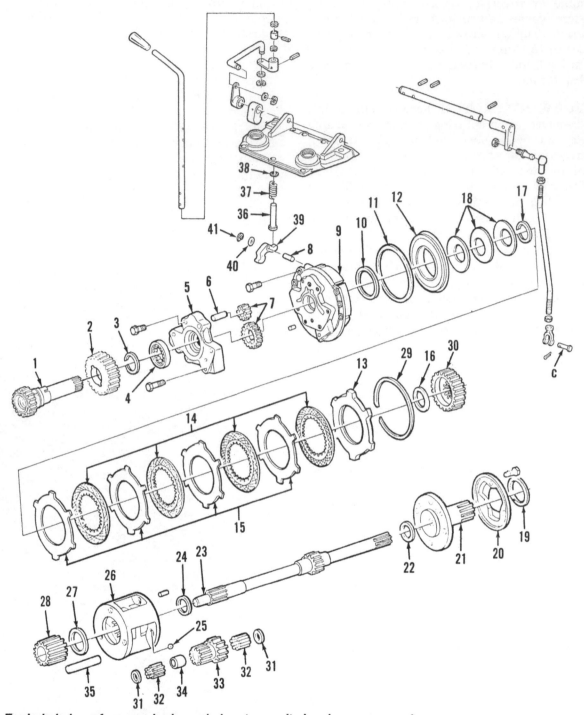

Fig. 114—Exploded view of reverse brake and planetary unit showing components.

1. Pto drive shaft	12. Brake piston	22. Thrust washer	32. Needle rollers
2. Pto drive gear	13. Backing plate	23. Clutch shaft	33. Planet pinion
3. Snap ring	14. Brake discs	24. Thrust washer	34. Spacer
4. Bearing	15. Separator plates	25. Steel ball	35. Planet pinion shaft
5. Transmission pump body	16. Thrust washer	26. Planet carrier	36. Pin
6. Dowel pin	17. Snap ring	27. Thrust washer	37. Spring
7. Pump gears	18. Belleville washers	28. Secondary sun gear	38. Washer
8. Pivot pin	19. Thrust washer	29. Snap ring	39. High speed lockout arm
9. Brake housing	20. Baffle	30. Brake hub	40. Washer
10. Sealing ring	21. Carrier shaft	31. Thrust washer	41. Snap ring
11. Piston ring			

so that piston (12) and seals (10 and 11) can be renewed. Refer to the following specifications.

Brake discs (14)—
  Thickness, new . . . . 2.84-3.00 mm
                               (0.112-0.118 inch)
    Wear limit . . . . 2.79 mm
                                  (0.110 inch)
  Groove depth, new . . . . 0.25-0.51 mm
                               (0.010-0.020 inch)
    Wear limit . . . . 0.20 mm
                                  (0.008 inch)

Separator plates (15)—
  Thickness, new . . . . 2.16-2.41 mm
                                (0.085-0.095 inch)

Belleville springs (18)—
  Free height, new . . . . 3.63 mm
                                  (0.143 inch)

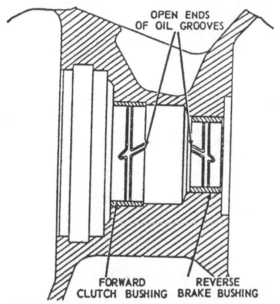

Fig. 114A—Cross section of clutch housing showing bushings properly installed with open ends of oil grooves toward center.

Fig. 114B—Reverse brake parts (36, 37 and 38) must be secured in housing bore before installing the transmission oil pump, reverse brake and planetary unit.

Minimum limit . . . . 3.12 mm
                                        (0.123 inch)

Assemble by reversing the removal procedure. Note the stamped numbers on large gear teeth of pinions (33) and corresponding numbers stamped on front of gear on clutch shaft (23). Time planetary unit during assembly by aligning corresponding numbers on clutch shaft and pinions.

Screws attaching carrier shaft (21) to carrier (26) should be tightened to 50 N·m (35 ft.-lbs.); screws attaching transmission pump body to brake housing should all be tightened to 30 N·m (23 ft.-lbs.) torque.

Bushings in clutch housing are renewable. If new bushings are installed, make sure open ends of oil grooves are together as shown in Fig. 114A. Before installing reverse brake, planetary and oil pump unit in clutch housing, be sure parts (39, 40 and 41—Fig. 114) are installed on pin (8) and parts (36, 37 and 38) are correctly positioned in clutch housing. Plunger pin (36) can be held up and in place as shown in Fig. 114B. Screws attaching brake housing to clutch housing should be tightened to 25 N·m (20 ft.-lbs.) torque.

115. BRAKE AND CLUTCH PISTONS. A special compressing tool such as JDT-24 must be used to compress Belleville washers (50—Fig. 112 and 18—Fig. 114) so snap ring can be removed and reinstalled. Square cut sealing rings (47 and 49—Fig. 112 and 10

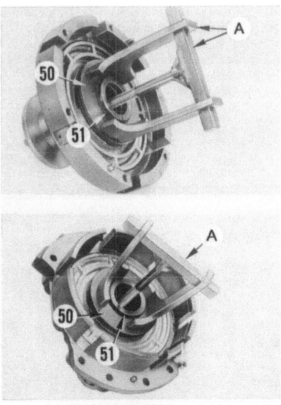

Fig. 115—Special tool should be used to compress Belleville washers (18 and 50).

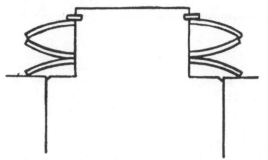

Fig. 115A—Belleville washers should be installed on clutch and brake pistons as shown to provide correct spring.

and 11—Fig. 114) are used on pistons. Lubricate seal rings thoroughly and be careful not to cut or otherwise damage sealing rings when assembling. Install Belleville washers as shown in Fig. 115A so that last installed spring washer has small inner diameter arched against snap ring.

CREEPER TRANSMISSION

OPERATION

Models So Equipped

116. The creeper transmission is a mechanically operated, synchronized reduction drive unit. The 2-speed unit operates as direct drive in Hi position in all transmission speeds and at a 79 percent reduction in Creeper position in Lo range I and reverse. A locking pin (15—Fig. 118) prevents engagement of Creeper drive in Hi range II.

The engine clutch must be disengaged before shifting Creeper transmission. Creeper transmission is located directly in front of the transmission in the clutch housing.

TROUBLE-SHOOTING

Models So Equipped

117. Some problems that may occur during operation of the Creeper transmission and their possible causes are as follows:

1. Noise during shifts. Could be caused by:
    a. Synchronizer discs worn.
    b. Engine clutch not fully disengaging.
    c. Shifter collar worn.

2. Noise during operation. Could be caused by:
    a. Taper roller bearings worn.
    b. Needle bearings worn.
    c. Incorrect preload of taper roller bearings.

3. Creeper gear jumps out of engagement. Could be caused by:
    a. Shifter fork worn.
    b. Shift linkage incorrectly adjusted.
    c. Shift linkage worn or damaged.

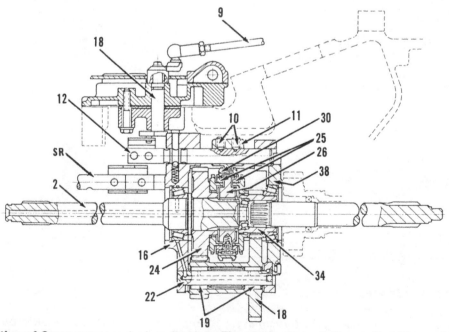

Fig. 116— Cross section of Creeper transmission. Refer to Fig. 118 for legend. Range II shifter rod (SR) is also shown.

R&R AND OVERHAUL

Models So Equipped

118. To remove the Creeper transmission, first remove the Sound Gard Body as outlined in paragraph 186, if so equipped, then split tractor between clutch housing and transmission as outlined in paragraph 127. Disconnect shift linkage, then unbolt and remove transmission shift cover. Refer to Fig. 117 and remove oil lines (4, 5 and 9) and sleeve (8). Remove cap screws (1) and lift out Creeper transmission.

NOTE: When removing Creeper transmission, be sure not to lose locking pin (15—Fig. 118), which may fall out.

To disassemble the unit, clamp transmission input shaft (2) in a vise with shaft pointing downward.

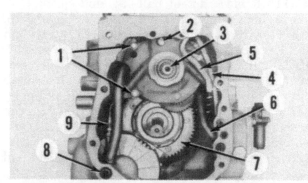

Fig. 117—Rear view of clutch housing showing Creeper transmission and component parts.

1. Creeper retaining cap screws
2. Shifter rod
3. Transmission input shaft
4. Lubricating line to oil cooler
5. Inlet line from filter
6. Oil pump pressure line
7. Pto clutch
8. Sleeve
9. Oil pump suction line

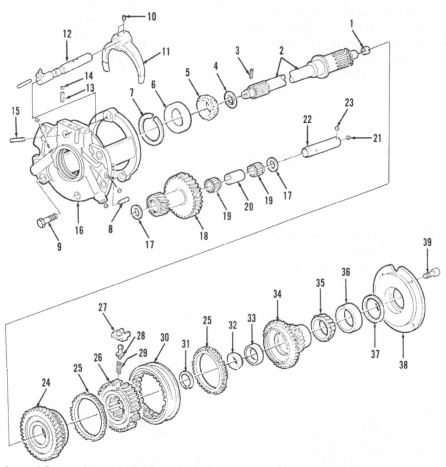

Fig. 118—Exploded view of Creeper transmission assembly.

1. Bushing (clutch shaft pilot)
2. Transmission input shaft
3. Spring pin
4. "O" ring
5. Bearing cone
6. Bearing cup
7. Snap ring
8. Dowel pin
9. Cap screw (4)
10. Set screws
11. Shift fork
12. Shifter rod
13. Spring
14. Detent ball
15. Locking pin
16. Carrier
17. Thrust washers
18. Cluster gear
19. Needle bearings
20. Spacer
21. Ball plug
22. Shaft
23. Lock ball
24. Creeper gear
25. Synchronizer discs
26. Synchronizer hub
27. Thrust block (3)
28. Ball pin (3)
29. Spring (3)
30. Shift collar
31. Snap ring
32. Bearing cone
33. Bearing cup
34. Input gear
35. Bearing cone
36. Bearing cup
37. Shim
38. End plate
39. Allen screw (4)

Loosen set screws (10) and remove shifter rod (12) and fork (11). Remove detent ball (14) and spring (13). Remove Allen screws (39) and lift off end plate (38). Remove bearing cup (36) and shims (37) from end plate. Remove shaft (22) taking care not to lose lock ball (23). Lift out cluster gear (18) with thrust washers (17), needle bearings (19) and spacer (20). Remove input gear (34), pull off bearing cone (35) and press out bearing cup (33). Pull bearing cone (32) from end of transmission shaft (2). Remove snap ring (31) and slide synchronizer assembly (25 through 30) and gear (24) from shaft (2). Remove unit from vise and pull shaft assembly from carrier (16). Discard "O" ring (4) and remove bearing cone (5) from shaft. Remove snap ring (7) and bearing cup (6) from carrier (16). Disassemble synchronizer unit.

Clean and inspect all parts for excessive wear or other damage and renew as necessary.

Reassemble by reversing disassembly procedure keeping the following points in mind: To adjust taper bearing preload, install bearing cup (36) without shims (37) in end plate (38). Install end plate and secure with Allen screws. Using a dial indicator check end play of input gear (34). This measurement plus 0.05 mm (0.002 inch) is the correct shim pack (37) to be installed. Remove end plate, press out bearing cup, install shim pack and reinstall bearing cup. Install end plate, apply Loctite 271 to threads of Allen screws and tighten to 25 N·m (20 ft.-lbs.) torque. Tighten set screws (10) to 40 N·m (30 ft.-lbs.) torque.

NOTE: Before installing Creeper transmission, coat locking pin (15) with grease and install in carrier (16). When joining tractor, make certain locking pin does not fall out.

Reinstall Creeper transmission assembly, tightening retaining screws to 50 N·m (35 ft.-lbs.) torque. Install oil lines (4, 5 and 9—Fig. 117) and sleeve (8).

Fig. 118A— Install synchronizer hub (26) so that thrust blocks (27) engage the third tooth (arrow) of shift collar as shown.

Install new "O" ring (4—Fig. 118) on transmission input shaft (2). Rejoin tractor and tighten cap screws to 160 N·m (120 ft.-lbs.) torque. Install transmission shift cover and connect shift linkage. Install Sound Gard Body, if so equipped.

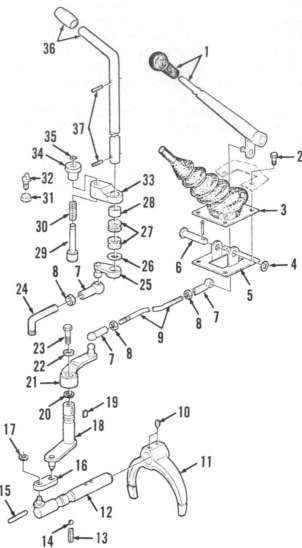

Fig. 119—Exploded view of Creeper transmission shift linkage. Some parts are also shown in Fig. 118.

1. Control lever		
2. Cap screw (4)	20.	"O" ring
3. Cover	21.	Arm
4. Washer	22.	Washer
5. Plate	23.	Cap screw
6. Pin	24.	Rod
7. Ball joint	25.	Arm
8. Jam nut	26.	Washer
9. Rod	27.	Bushings
10. Set screw (2)	28.	Sleeve
11. Shift fork	29.	Pin
12. Shifter rod	30.	Spring
13. Spring	31.	Nut
14. Detent ball	32.	Screw
15. Locking pin	33.	Link
16. Link	34.	Knob
17. Lock ring	35.	Snap ring
18. Shifter shaft	36.	Control lever
19. Key	37.	Spring pins

TRANSMISSION
(COLLAR SHIFT)

OPERATION

119. The transmission used on some models is a collar shift unit with helical cut gears. The two levers are located on top of the clutch housing in shift cover. The left lever selects high or low range and reverse on models without direction reverser. The right lever selects one of four stepped gears. These shift levers (plus the Direction Reverser shift unit on models so equipped) combine to make possible 8 forward and 4 reverse gears.

Disengage engine clutch when shifting transmission gears.

TOP (SHIFT) COVER

120. REMOVE AND REINSTALL. To remove the shifter cover (1—Fig. 120), first remove transmission shield. Disconnect the hydraulic oil reservoir leak-off and breather line from cover. Detach wire from transmission oil pressure warning switch, then lay wire away from cover. Disconnect Hi-Lo clutch linkage, then unbolt and remove control cover from shift cover. Remove attaching screws and lift shift cover with shift levers from housing. Any further disassembly will be obvious after an examination of the unit and reference to Fig. 120.

Reinstall by reversing the removal procedure.

NOTE: Shifter shafts and forks are an integral part of the transmission and can be serviced after transmission is split from clutch housing as outlined in paragraph 121.

TRACTOR SPLIT

121. To split tractor between clutch housing and transmission case, first drain transmission case and remove the top shift cover as outlined in paragraph 120. Remove hydraulic oil filter cover and element, then disconnect all wires and tubes that would interfere when separating. Support rear of tractor under transmission case and front of tractor under clutch housing.

CAUTION: Be sure tractor is supported adequately to permit separation and will not tip or allow other unwanted movement.

Remove the two clutch housing to transmission case cap screws at rear of shift cover opening under the mounting flange. Remove the remaining nine screws, then separate clutch housing and transmission case.

NOTE: Do not lose the pto shaft coupling. If the pto coupling falls off shaft and into transmission case, use screwdriver to reposition coupling on rear shaft before attempting to reattach clutch housing and transmission case.

When reassembling, move clutch housing and transmission case together completely before installing attaching cap screws. Coat threads of screw with sealer before installing in hole third from top on right. Remainder of assembly is reverse of disassembly procedure. Tighten clutch housing to transmission case cap screws to a torque of 160 N·m (120 ft.-lbs.).

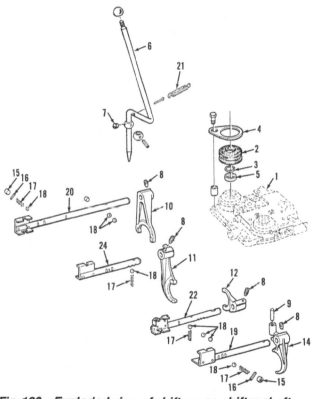

Fig. 120—Exploded view of shift cover, shifter shafts and shift forks used.

1. Shift cover		14.	Fork (Range I-Reverse)
2. Boot		15.	Plug
3. Snap ring		16.	Spring pins
4. Retainer		17.	Detent springs
5. Bushing		18.	Detent balls
6. Shift lever		19.	Shifter shaft (Range I-Reverse)
7. Set screw		20.	Shift shaft (1-5 & 2-6)
8. Set screws		21.	Spring
9. Start safety switch pin		22.	Shifter shaft (Range II)
10. Fork (1-5 & 2-6)		24.	Shift shaft (3-7 & 4-8)
11. Fork (3-7 & 4-8)			
12. Fork (Range II)			

SHIFTER SHAFTS AND FORKS

122. REMOVE AND REINSTALL. To remove shifter shafts and forks, first split tractor as outlined in paragraph 121. Remove seat and disconnect lift links from rockshaft arms. Detach rear wiring harness and disconnect lines from selective control valves to rockshaft housing. Attach a hoist to rockshaft housing, place load selector lever in "L" position, then unbolt and remove rockshaft housing from transmission case.

Remove starter safety switch pin (9—Fig. 120) and set screws (8) from shift forks (12 and 14).

Move both range shifter shafts (19 and 22—Fig. 122) to neutral position, then pull shaft (19) out of case and shift fork (14). **Do not turn shifter shafts while withdrawing from housing. Detent ball could drop into set screw hole, making shaft impossible to remove. Also, do not lose interlock balls, detent balls or springs when shifter shafts are removed from case bores.** Pull shaft

(22) from case and shift fork (12). Remove interlock balls, detent balls and springs.

Move shifter shafts (20 and 24) to neutral position, remove set screws from shift forks (10 and 11), then withdraw shifter shaft (24) from case bore. Carefully remove interlock balls, detent balls and springs. Remove shifter shaft (20). Remove forks (10 and 11) as shafts are withdrawn.

Inspect all parts for evidence of wear, rust or other damage and renew as necessary. Assemble shifter shafts and forks in reverse of order in which they were removed.

COUNTERSHAFT

123. R&R AND OVERHAUL. To remove the countershaft, split tractor as outlined in paragraph 121 and remove shift forks and shaft as outlined in paragraph 122. Remove the pto gear, or gears, from front of transmission. Remove cap screws from countershaft bearing support (13—Fig. 124). Remove snap ring (33) from its groove at rear end of countershaft, then use a screwdriver and turn lock washer (29) until splines of washer index with splines of countershaft. Pry bearing support off dowels, pull assembly forward, notice that shoulder on shift collar (28) is toward front and lift gears from transmission case as they come off shaft.

Inspect all gears, thrust washers and shift collar for broken teeth, excessive wear or other damage and renew as necessary. If support assembly bearings or shafts require service, the shafts and bearings can be pressed out after removing snap rings; however, the transmission drive gear must be removed before countershaft can be removed. Snubber brake springs (22) should test 280-350 N (63-79 lbs.) when compressed to a height of 38.5 mm (1.51 inches). Needle bearing (35) can be removed from its bore after removing snap ring (34).

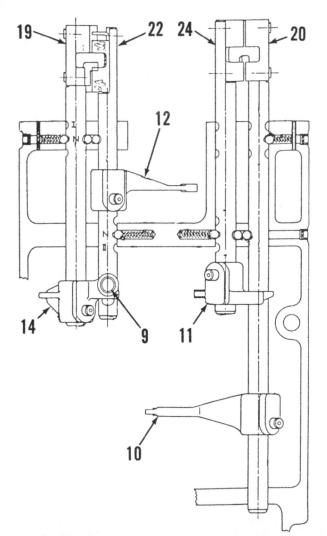

Fig. 122—Drawing of shift forks, shifter shafts and associated parts. Refer to Fig. 120 for legend.

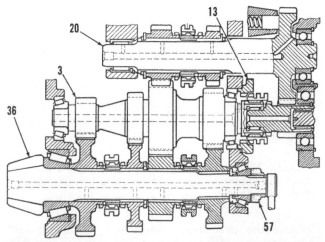

Fig. 123—Cross section of transmission shafts, gears and collars. Refer to Fig. 124 for legend.

On models with reverser, plug (20) is used to plug hole in countershaft (24) where reverse pinion (27) was installed on models without reverser. Also, brake snubber spring shims (23) are used to increase spring pressure behind snubber brake plugs (21).

INPUT SHAFT

123A. R&R AND OVERHAUL. To remove the input shaft it is first necessary to remove the countershaft as outlined in paragraph 123.

With countershaft removed, the transmission input shaft is removed as follows: Remove transmission oil cup and lines. Straighten lock plates and remove input shaft bearing quill (8—Fig. 124) and shims (7) from front of input shaft. Bump input shaft forward, lift rear end of shaft and remove input shaft from transmission case.

With input shaft removed, inspect all gears for chipped teeth or excessive wear. Inspect bearings and renew as necessary. Bump bearing cup (1) forward if removal is required. Inspect needle bearing (4) and renew if necessary.

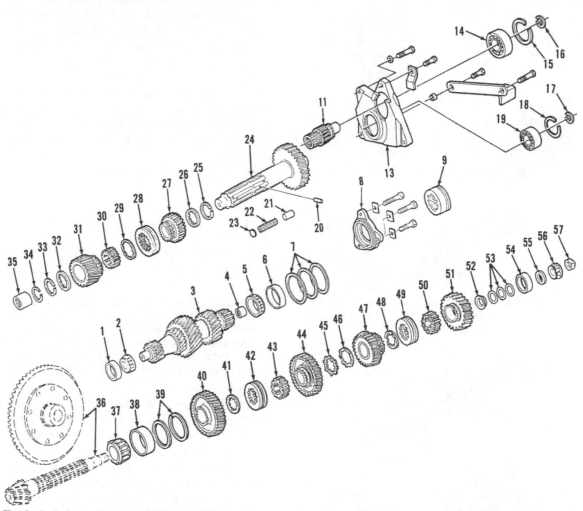

Fig. 124—Exploded view of transmission shafts, gears, shift collars and associated parts used on collar shift units. Part (27) is not used with Reverser and plug (20) is used.

1. Bearing cup	17. Snap ring	31. Low range pinion
2. Bearing cone	18. Snap ring	32. Thrust washer
3. Input shaft	19. Ball bearing	33. Snap ring
4. Needle bearing	20. Plug	34. Snap ring
5. Bearing cone	21. Brake plug	35. Needle bearing
6. Bearing cup	22. Spring	36. Pinion shaft
7. Shims	23. Shim	37. Bearing cone
8. Bearing quill	24. Countershaft	38. Bearing cup
9. Shifter collar	25. Snap ring	39. Shims
11. Drive gear	26. Thrust washer	40. 1st & 5th gear
13. Support	27. Reverse pinion	41. Thrust washer
14. Ball bearing	28. Shift collar	42. Shift collar
15. Snap ring	29. Thrust washer (locking)	43. Shift collar sleeve
16. Snap ring	30. Shift collar sleeve	44. 2nd & 6th gear

45. Thrust washer (outer tangs)	
46. Retaining washer	
47. 4th & 8th gear	
48. Thrust washer	
49. Shift collar	
50. Shift collar sleeve	
51. 3rd & 7th gear	
52. Spacer	
53. Shims	
54. Bearing cup	
55. Special washer	
56. Bearing cone	
57. Nut	

Install and adjust end play of input shaft as follows: Be sure bearing cup (1) is bottomed in bore and place input shaft in position. Use original shim pack (7), or use a shim pack approximately 0.8 mm (0.030 inch) thick, install front bearing quill (8) and tighten retaining cap screws to 50 N·m (35 ft.-lbs.) torque. Use a dial indicator to check input shaft end play, which should be 0.10-0.15 mm (0.004-0.006 inch). Vary

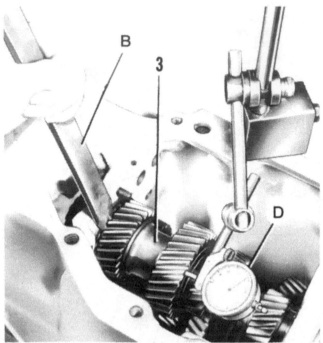

Fig. 127—Transmission drive shaft end play can be measured with dial indicator (D) as shown. Refer to text.

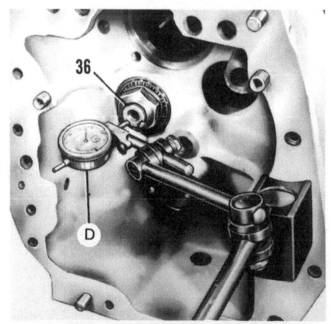

Fig. 128—Use a bar to move countershaft while measuring countershaft end play with dial indicator (D).

thickness of shims (7) as required. Do not forget front oil line clamp when making final assembly.

PINION SHAFT

123B. R&R AND OVERHAUL. To remove the transmission pinion shaft, first remove the shift forks and shafts as outlined in paragraph 122, the countershaft as outlined in paragraph 123, the input shaft as outlined in paragraph 123A, both final drive assemblies as outlined in paragraph 139 or 147 and the differential as in paragraph 133.

Remove oil line, nut (57—Fig. 124), bearing (56), shims (53) and spacer (52). Use screwdriver and turn thrust washers (41, 45 and 48) until splines of thrust washers are indexed with splines of pinion shaft. Pull pinion shaft rearward and remove parts from transmission case as they come off shaft. Bearing cup (38) and shims (39) can be removed from housing by bumping cup rearward. Be sure to keep shims (39) together as they control the bevel gear mesh position. Bearing cup (54) can be removed by bumping cup forward.

Check all gears and shafts for chipped teeth, damaged splines, excessive wear or other damage and renew as necessary. If pinion shaft is renewed, it will also be necessary to renew the differential ring gear and right hand differential housing as these parts are not available separately. Bearing (37) is installed with large diameter toward gear end of shaft.

NOTE: Mesh (cone point) position of the pinion shaft and main drive bevel pinion gear is adjusted with shims (39) located between rear bearing cup (38) and housing. If new drive gears or bearings are installed, the mesh position must first be checked and adjusted as outlined in paragraph 136.

Install pinion shaft and adjust shaft bearing preload as follows: Use Fig. 124 as a guide and with bearing (37) on pinion shaft, start shaft into rear of housing. With shaft about half-way into housing, place 1st and 5th speed gear (40) on shaft with teeth for shift collar (42) toward front. Place thrust washer (41) on shaft, then install coupling sleeve (43) and shift collar (42). Move shaft forward slightly and install 2nd and 6th speed gear (44) with teeth for shift collar toward rear. Place thrust washer with outer tangs (45) over shaft, then slide retaining washer (46) over thrust washer (45). Move shaft slightly forward and install 4th and 8th speed gear (47) on shaft with teeth for shift collar toward front. Place thrust washer (48) on shaft an install shift collar sleeve (50) and shift collar (49). Install 3rd and 7th speed gear (51) on shaft with teeth for shift collar toward rear. Push shaft forward until rear bearing cone (37) seats in bearing cup (38) and use screwdriver to turn thrust washer until splines on thrust washers lock with splines of pinion shaft. Install spacer (52), shims (53), bearing (56), washer (55) with tang facing forward, and nut (57), then adjust pinion shaft bearing preload as outlined in paragraph 123C.

123C. PINION SHAFT BEARING ADJUSTMENT. The pinion shaft bearings must be adjusted to provide a bearing preload of 0.15 mm (0.006 inch). Adjustment is made by varying the number of shims (53—Fig. 124).

To adjust the pinion shaft bearing preload, proceed as follows: Mount a dial indicator with contact button on front end of pinion shaft and check for end play of shaft. If shaft has no end play, add shims (53) to introduce not more than 0.05 mm (0.002 inch) shaft end play.

NOTE: Do not exceed 0.05 mm (0.002 inch) shaft end play when beginning adjustment because more end play increases the probability of inaccurate measurements and resulting improper adjustment of an important bearing preload adjustment.

If original shims (53) are not being used, install preliminary 0.85 mm (0.035 inch) thickness shim pack. Shims are available in 0.05, 0.13, 0.25 and 0.80 mm (0.002, 0.005, 0.010 and 0.032 inch) thicknesses. Tighten nut (57) to 220 N·m (160 ft.-lbs.) torque and measure shaft end play. Remove shims equal to the measured end play PLUS and additional 0.15 mm

(0.006 inch). This will give the recommended bearing preload of 0.15 mm (0.006 inch). Retighten nut (57) to 220 N·m (160 ft.-lbs.) torque and stake in position.

TRANSMISSION OIL PUMP

124. The transmission oil pump is an external gear-type pump mounted on rear wall of clutch housing as shown at (7—Fig. 114) for models with direction reverser or Fig. 130 for models without. Oil from the transmission oil pump is delivered to the oil cooler, to tractor brakes and to the lubricating systems for the reverser and collar shift transmissions. The transmission oil pump supplies pressurized oil to operate the reverser transmission and also delivers oil to the main hydraulic pump.

A bypass valve (1 through 6—Fig. 129) is located above the filter in transmission housing. If the filter (16) becomes clogged or if oil is too cold to circulate through filter easily, the differential pressure (before the filter and after the filter) will exceed the opening pressure of the bypass valve and oil will be directed back to the main oil reservoir (transmission case). Differential pressure should be 200-350 kPa (28.5-50 psi). Specifications for spring (5) should be as follows:

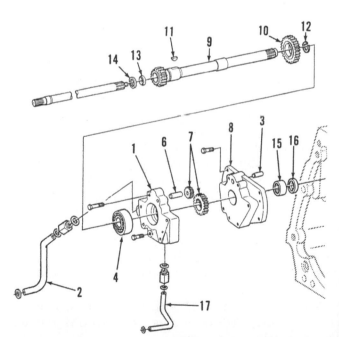

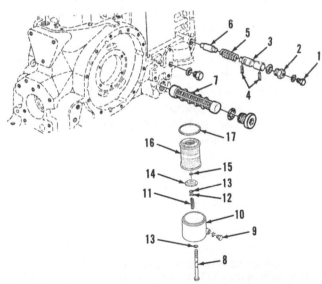

Fig. 129—View showing locations of transmission oil pump screen (7), filter (16) and by-pass valve (3).

1. Plug	
2. Threaded nipple	10. Filter cover
3. By-pass valve	11. Compression spring
4. Spring pins	12. Washer
5. Spring	13. "O" ring
6. Pin	14. Retainer
7. Inlet screen	15. Snap ring
8. Bolt	16. Filter
9. Plug	17. Seal washer

Fig. 130—Exploded view of transmission oil pump used on models with collar shift transmission but without reverser.

1. Pump housing	10. Gear
2. Suction oil tube	11. Drive key
3. Dowel pin	12. Snap ring
4. Roller bearing	13. Cap
6. Idler shaft	14. Seal
7. Pump gears	15. Needle bearing
8. Pump adapter	16. Seal
9. Pto drive shaft	17. Pressure tube

Models 2150, 2350, 2550
Free length .... Approx. 68 mm
(2.68 inches)
Pressure when compressed to: .. 49 mm/50-60 N
(1.5 inches/11-13 lbs.)
Pressure when compressed to: .. 38 mm/80-95 N
(1.93 inches/18-21 lbs.)

Models 2155, 2255, 2355, 2355N, 2555
Free length .... Approx. 65 mm
(2.56 inches)
Pressure when compressed to 33 mm .... 42-52 N
(1.3 inches. 9.5-11.5 lbs.)

If volume of flow is suspected to be low, remove, clean and inspect transmission pump inlet screen (7).

125. R&R AND OVERHAUL. To remove the transmission oil pump, first split transmission from clutch housing as outlined in paragraph 121.

With clutch housing separated from transmission case, pull pto clutch from clutch housing. Remove pump inlet and outlet lines from pump, then remove pump from wall of clutch housing. Separate pump body from adapter.

Clean and inspect all parts for chipping, scoring or excessive wear. If bearing in pump body requires renewal, press new bearing in bore until it bottoms. Pump gears are available as a matched set only. Pump idler shaft is renewable and diameter of new shaft is 15.85-15.87 mm (0.624-0.625 inch). Idler gear bushing ID is 15.91-15.94 mm (0.626-0.627 inch). Thickness of new pump gears is 12.90-12.93 mm (0.508-0.509 inch). Clearance between pump housing and gears should be 0.1-0.2 mm (0.0004-0.0008 inch).

When reassembling pump, coat gears with oil and tighten adapter mounting cap screws and housing to adapter cap screws to 50 N·m (35 ft.-lbs.) torque. Align slots of pto powershaft with lugs of pump drive gear when installing shafts. Make certain seals are on ends of inlet and outlet tubes before rejoining tractor.

TRANSMISSION (SYNCHRONIZED)

OPERATION

126. The transmission used on SOME Models is of the cone synchronized type. Tractors without Sound Gard Body are equipped with center shift (two shift levers in shift cover). Tractors with Sound Gard Body are equipped with console shift (two shift levers in console to right of operator's seat). The range shift lever (left lever on center shift or right lever on console shift) is used to select low range I, high range II or reverse range. The gear shift lever (right lever on center shift or left lever on console shift) is used to select the four synchronized gears of each range. The transmission, when combined with the Hi-Lo shift unit, provides a total of 16 forward gears and eight reverse gears.

Tractors are equipped with a neutral start safety switch. Range shift lever must be in neutral position before engine can be started.

TRACTOR SPLIT

127. To split tractor between clutch housing and transmission case, drain transmission and disconnect battery cables. On models so equipped, remove Sound Gard Body as outlined in paragraph 186. On models equipped with front-wheel drive, disconnect front drive shaft. If so equipped, disconnect Hi-Lo or Creeper shift linkage. On all models, disconnect hydraulic reservoir leak-off and breather line from shifter cover. Detach wire from transmission oil pressure warning switch. Remove attaching cap screws and remove shifter cover assembly. Disconnect all wires and hydraulic tubes that would interfere when separating tractor. Support rear of tractor under transmission case and front of tractor under clutch housing. Place wood block between front axle and front support to prevent tipping. Remove the two clutch housing to transmission case cap screws located at rear of shifter cover opening. Remove the remaining cap screws, then separate clutch housing and transmission case.

NOTE: On models equipped with Creeper transmission, be careful not to lose locking pin (15—Fig. 118), which may fall out when splitting tractor. Coat locking pin with grease and install in carrier (16) when rejoining tractor.

Rejoin tractor and tighten cap screws to a torque of 160 N·m (120 ft.-lbs.). Balance of reassembly is the reverse of disassembly. Fill transmission to top mark on dip stick with John Deere Hy-Gard or Quatrol oil or equivalent. On models so equipped, install Sound Gard Body.

SHIFT LINKAGE, SHIFTER SHAFTS AND FORKS

Models With Center Shift

128. REMOVE AND REINSTALL. To remove the shift levers (2—Fig. 131) and shift cover (9), first remove transmission shield by working it up over shift levers. Disconnect Hi-Lo or Creeper shift linkage, if so equipped. Disconnect rear wiring harness, then unbolt and remove shift cover assembly. To remove shifter shafts and shift forks, first split tractor between clutch housing and transmission case as outlined in paragraph 127. Remove rockshaft assembly as in paragraph 177. Remove safety start switch pin (12—Fig. 131). Remove set screws (10) from shift forks and guide. Place shifter shafts in neutral posi-

tion, then pull shifter shafts one at a time forward and remove. Be careful to catch detent and interlock balls as shafts are withdrawn.

Remove shift forks and guide. To remove low range I and reverse shift fork, it is necessary to remove transmission bearing quill.

Clean and inspect all parts and renew any showing excessive wear or other damage.

Reassemble by reversing disassembly procedure. Make certain detent balls (17—Fig. 131), springs (18) and interlock balls (19) are installed correctly. Adjust position of shift forks before set screws (10) are fully tightened. Shift forks must be positioned on shifter shafts so that shift forks do not rub against shifter collar. Set position of shift rails to neutral detent position and tighten set screws lightly. Check position of shift forks in synchronizer collars and make sure all are centered in neutral. After forks are correctly

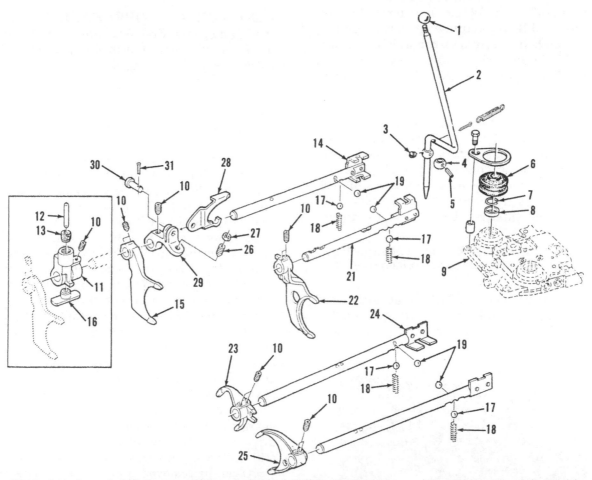

Fig. 131—Exploded view of shift levers, shifter shafts and shift forks used on models with center shift. Guide parts (28, 29, 30 and 31) are used on later models instead of parts (11, 12, 13 and 16).

1. Knob	9. Shift cover	17. Detent ball (4)	25. Shift fork (1-5 & 2-6)
2. Shift lever (2)	10. Set screws	18. Spring (4)	26. Set screw
3. Special set screw	11. Guide	19. Interlock ball (4)	27. Lock nut
4. Ball	12. Safety start switch pin	21. Shifter shaft (low range I & reverse)	28. Adjusting screw
5. Spring pin	13. Pin bushing	22. Shift fork (low range I & reverse)	29. Guide
6. Boot	14. Shifter shaft (high range II)	23. Shift fork (3-7 & 4-8)	30. Pin
7. Snap ring	15. Shift fork (high range II)	24. Shifter shaft (3-7 & 4-8)	31. Cotter pin
8. Bearing	16. Shoe		

adjusted, tighten set screws evenly to a torque of 40 N•m (30 ft.-lbs.). Make sure forks do not rub against any part of collar.

Reinstall rockshaft assembly and rejoin tractor. Install shift cover and transmission shield.

Models With Console Shift

129. REMOVE AND REINSTALL. To remove transmission shift linkage, refer to Fig. 132 and remove cap screws (4) and rockshaft lever stop nut (5). Remove console side panel (3). Remove knobs from hand throttle, selective control valve, rockshaft control lever, gear shift lever and range shift lever. Disconnect wire at Hi-Lo shift unit indicator light. Unbolt console from fender and lift console (6) upward and remove. Disconnect throttle rod, Hi-Lo connecting rod, if so equipped and shift rods (14, 15, 16 and 17—Fig. 133). Unbolt and remove bracket (4) with shift lever assemblies. Disconnect and remove shift rods from lever shafts (18, 19, 20 and 21).

To remove shifter shafts and shift forks, first remove Sound Gard Body as outlined in paragraph 186, separate clutch housing from transmission case as in paragraph 127 and remove rockshaft assembly as in paragraph 177. Move all shifter shafts to neutral position. Mark position of shift arms (28, 29, 35 and 36) in relation to lever shafts (18, 19, 20 and 21). Loosen set screws in shift arms, pull lever shafts from housing (24) and remove shift arms. Loosen set screws in shift forks, safety start switch arm (33) and guide (38). Pull shifter shafts (34 and 37), one at a time, out front of transmission and remove shift fork (32), safety start switch arm (33) and guide (38). Catch detent and interlock balls as shafts are removed.

> **CAUTION: Do not turn shifter shafts when pulling them forward as detent balls and springs may fall out.**

Pull shifter shafts (27 and 31), one at a time, out front of transmission and remove forks (26 and 30). To remove shift fork (39), it is necessary to remove

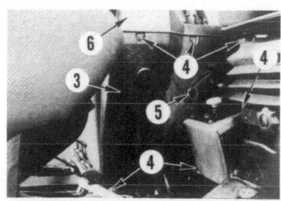

Fig. 132—View of console typical of all models equipped with Sound Gard Body.

3. Side panel	5. Rockshaft lever stop
4. Cap screws	6. Console cover

bearing quill from front of transmission. Unbolt and remove housing (24—Fig. 133).

Clean and inspect all parts and renew any showing excessive wear or other damage.

Reassemble by reversing disassembly procedure. Make certain detent balls (41), springs (42) and interlock balls are installed correctly. Adjust position of shift forks before set screws are fully tightened. Shift forks must be positioned on shifter shafts so that shift forks do not rub against shifter collar. Set position of shift rails to neutral detent position and tighten set screws lightly. Check position of shift forks in synchronizer collars and make sure all are centered in neutral. After forks are correctly adjusted, tighten set screws evenly to a torque of 40 N•m (30 ft.-lbs.). Make sure forks do not rub against any part of collar.

Reinstall rockshaft assembly and rejoin tractor. Install Sound Gard Body.

SHAFTS AND GEARS

130. R&R AND OVERHAUL. To remove the shafts and gears, first separate clutch housing from transmission case as outlined in paragraph 127. Remove rockshaft housing assembly as in paragraph 177 and shifting mechanism as in paragraph 128 or 129. Refer to Fig. 135, straighten locking tab and remove special nut (38). Remove cap screws (A), then remove bearing quill with reverse idler and if equipped with front-wheel drive, intermediate shaft (84). Remove shaft spacer (34—Fig. 134), shim (35) and gear (30B), if so equipped. Remove snap ring (33—Fig. 135A) from early models or screws and retainer plate (33A) from later models. Pull extension shaft (74—Fig. 136) with shift fork (A) off bevel pinion shaft. Remove oil cup with lube oil lines. Unbolt and remove bearing cap (B—Fig. 137), then remove snap ring (1—Fig. 134) from housing groove. Lift top shaft assembly (A—Fig. 137) from top opening of transmission case. Remove differential assembly as outlined in paragraph 133. Remove special nut (58 or 58A—Fig. 134), drive bevel pinion shaft (44) rearward and pull bearing cone (57) from shaft. Remove shim (55) and spacer (54). Pull bevel pinion shaft (44) out rear of transmission case, taking care not to damage needle bearings (51 and 53) in countershaft cluster gear. Remove countershaft cluster gear (52) with needle bearings, thrust washer (50), shift collar (49) and drive sleeve (48). If necessary, remove all bearing cups still located in transmission case. Needle bearing race (53A) can be removed if new race is to be installed. Heat new race before installing.

To disassemble transmission bearing quill, remove cap screw then remove shaft, reverse idler gear and bearing assembly.

On tractors without front-wheel drive, remove snap ring from groove in pto connector sleeve (83) and pull sleeve out of needle bearing (81). Remove snap rings, then push needle bearing (81) from bearing quill.

On tractors equipped with front-wheel drive, straighten lock ring (95) and remove special nut (96). Drive intermediate shaft (84) inward and remove bearing cone (94). Pull intermediate shaft out of bearing quill. Remove gear (89), thrust rings (88 and 91) and shims (90). Remove snap ring (80) and pull pto connecting sleeve (83) with needle bearing (81) out of intermediate shaft (84). Remove snap ring from connecting sleeve, then remove needle bearing from sleeve. Remove bearing cone (85) from intermediate shaft. If necessary, remove snap rings (87 and 92) and press bearing cups (86 and 93) from bearing quill.

To disassemble the top shaft assembly, separate low range I-reverse drive shaft (30) from drive shaft (19). Remove snap ring (28), thrust ring (28B) and needle

bearing (29) from shaft (30), then pull bearing cone (31) from shaft. Remove snap ring (27) from drive shaft (19) and slide synchronizer (20, 21, 22, 24, 25 and 26) and gear (23) from drive shaft. Pull bearing cone (3) off shaft, then remove snap rings (4) and "O" ring (5). Pull bearing cone (6) from shaft, remove snap ring (8), thrust ring (9), gear (10) and snap ring (11). Slide synchronizer assembly (12 through 17) and gear (18) from drive shaft (19). Some parts of synchronizers may be alike, but should be identified for installation in the same location.

Disassemble extension shaft as follows: On models without front-wheel drive, remove snap ring (67) and pull bearing cone (68) off extension shaft (59). Remove thrust washer (66) and gear (65) from extension shaft.

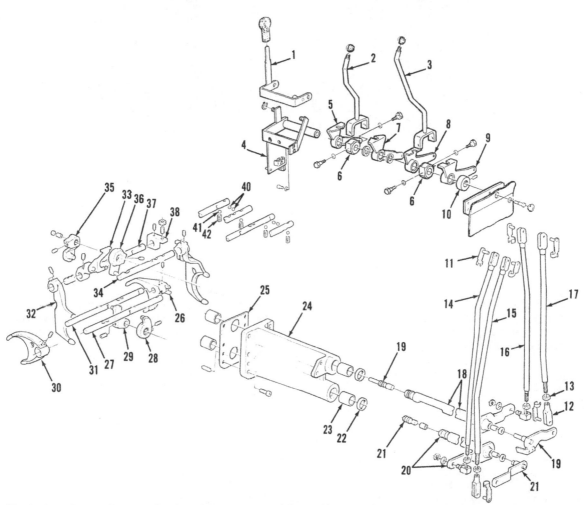

Fig. 133—Exploded view of transmission shift linkage, shifter shafts and shift forks used on models equipped with console shift.

1. Console shift lever	12. Yoke	23. Bushing (4)	33. Safety start switch arm
2. Gear shift lever	13. Locknut	24. Housing	34. Shifter shaft (low range I & reverse)
3. Range shift lever	14. Rod (1-5 & 2-6)	25. Gasket	35. Shift arm
4. Mounting bracket	15. Rod (3-7 & 4-8)	26. Shift fork (3-7 & 4-8)	36. Shift arm
5. Shift lever	16. Rod (low range I & reverse)	27. Shifter shaft (1-5 & 2-6)	37. Shifter shaft (high range II)
6. Bushing	17. Rod (high range II)	28. Shift arm	38. Guide
7. Shift lever	18. Hollow lever shaft (upper)	29. Shift arm	39. Shift fork (low range I & reverse)
8. Shift lever	19. Lever shaft (upper)	30. Shift fork (1-5 & 2-6)	40. Interlock balls (4)
9. Shift lever	20. Hollow lever shaft (lower)	31. Shifter shaft (3-7 & 4-8)	41. Detent ball (4)
10. Locating ring	21. Lever shaft (lower)	32. Shift fork (high range II)	42. Spring (4)
11. Clip pin	22. Seal ring (2)		

Remove snap ring (64), shift collar (63), shift sleeve (62), snap ring (61) and gear (60) from shaft (59 or 59A).

On models equipped with front-wheel drive, remove snap ring (77) and pull bearing cone (76) from extension shaft (74 or 74A). Remove gears (75 and

65), snap ring (64), shift collar (63), shift sleeve (62), snap ring (61) and gear (60) from shaft (74 or 74A).

Clean and inspect all parts and renew any showing excessive wear or other damage. Later extension shafts (59A and 74A) have lugs that lock position of nut (58A) when assembling.

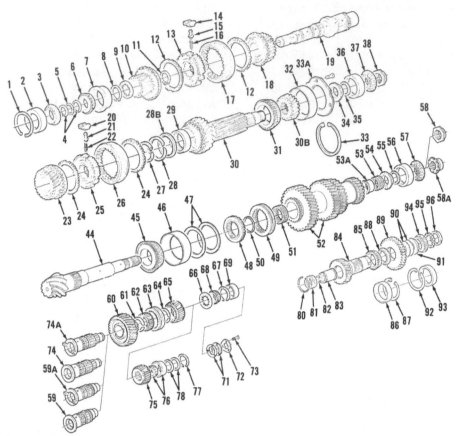

Fig. 134—Exploded view of transmission shafts, gears and associated parts typical of all models with synchronized transmission. Items (74 through 78 and 84 through 96) are used only on models equipped with front-wheel drive and parts with "A" suffix are used on later models. Removable reverse gear (30B) is not used on all models and reverse gear is machined as part of drive shaft (30).

1. Snap ring	27. Snap ring	53. Needle bearing	74. Extension shaft
2. Bearing cup	28. Snap ring	53A. Needle bearing race	(early MFWD)
3. Bearing cone	28B. Thrust ring	54. Spacer	74A. Extension shaft
4. Snap rings	29. Needle bearing	55. Shim	(late MFWD)
5. "O" ring	30. Drive shaft (low range	56. Bearing cup	75. Gear (front-wheel drive)
6. Bearing cone	I-reverse & 4-8)	57. Bearing cone	76. Taper roller bearing
7. Bearing cup	30B. Reverse gear	58. Special nut	77. Snap ring
8. Snap ring	(some models)	58A. Special nut	78. Shims
9. Thrust washer	31. Bearing cone	59. Extension shaft	80. Snap ring
10. Gear (1-5)	32. Bearing cup	(early models)	81. Needle bearing
11. Snap ring	33. Snap ring	59A. Extension shaft	82. Snap ring
12. Synchronizer ring	33A. Bearing retainer	(late models)	83. Pto connecting sleeve
13. Synchronizer hub	34. Spacer	60. Gear (low range I)	84. Intermediate shaft
14. Thrust block	35. Shim	61. Snap ring	85. Bearing cone
15. Ball pin	36. Bearing cup	62. Shift sleeve	86. Bearing cup
16. Spring	37. Bearing cone	63. Shift collar	87. Snap ring
17. Shift collar	38. Nut	64. Snap ring	88. Thrust ring
18. Gear (2-6)	44. Bevel pinion shaft	65. Gear (reverse range)	89. Intermediate gear
19. Drive shaft	45. Bearing cone	66. Thrust washer	90. Shim
20. Thrust block	46. Bearing cup	67. Snap ring	91. Thrust ring
21. Ball pin	47. Shim	68. Bearing cone	92. Snap ring
22. Spring	48. Drive sleeve (range II)	69. Bearing cup	93. Bearing cup
23. Gear (3-7)	49. Shift collar	71. Shim	94. Bearing cone
24. Synchronizer ring	50. Thrust washer	72. Plate	95. Lock ring
25. Synchronizer hub	51. Needle bearing	73. Cap screw (3)	96. Special nut
26. Shift collar	52. Countershaft cluster gear		

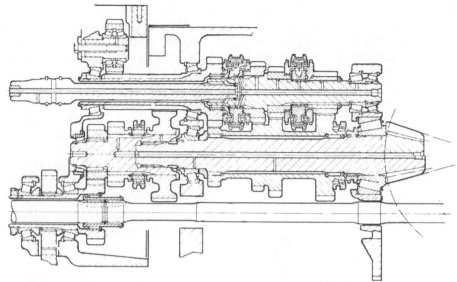

Fig. 134A—Cross sectional view of synchronized transmission. Refer to text.

NOTE: If bevel pinion shaft is worn or damaged and must be renewed, also renew ring gear and housing assembly as these parts are a matched set.

Reassemble transmission by reversing disassembly procedure and keeping the following points in mind: When installing reverse idler gear and shaft assembly in bearing quill, tighten cap screw to a torque of 55 N·m (40 ft.-lbs.).

If equipped with front-wheel drive, install intermediate shaft (84) and component parts (86 through 96) in bearing quill, using a shim pack (90) 1.15 mm (0.045 inch) thick. Tighten special nut to a torque of 140 N·m (100 ft.-lbs.). Using a dial indicator, measure intermediate shaft end play. This measurement plus

0.05-0.10 mm (0.002-0.004 inch) is the correct thickness of shims (90) to be removed to obtain correct preload.

To assemble drive shaft assembly (8 through 27), clamp drive shaft (19) in a protected vise so that internal splined bore is downward. Install 2-6 gear (18) and synchronizer ring (12) on shaft. Slide shift sleeve (13) on shaft with flat side facing downward. Select and install snap ring (11) of correct thickness to provide shift sleeve end play of 0-0.1 mm (0-0.004 inch). Snap rings (8 and 11) are available in thickness

Fig. 135—Front bearing quill installed on early model equipped with front-wheel drive.

A.	Cap screws
30.	Low range I-reverse
	drive shaft
38.	Special nut

43.	Reverse idler cap screw
74.	Extension shaft
84.	Intermediate shaft

Fig. 135A—Early models are equipped with snap ring (33) to retain bearing cup (32) and late models use Torx screws to retain plate (33A).

of 1.45, 1.55, 1.65 and 1.75 mm (0.057, 0.061, 0.065 and 0.069 inch). Install springs (16), ball pins (15) and thrust blocks (14), then slide shift collar into place so that third tooth beside each tooth gap is in mesh with corresponding thrust block (Arrow—Fig. 139). Install second synchronizer ring (12—Fig. 134), 1-5 gear (10) and thrust washer (9). Select and install snap ring (8) of correct thickness to provide 1-5 gear (10) end play of 0.0-0.4 mm (0.0-0.016 inch). Invert drive shaft (19) in vise and slide 3-7 gear (23) into position on drive shaft. Install second synchronizer in same manner as the first one. Make certain lube groove in shift sleeve (25) is aligned with lube bores in drive shaft. Select

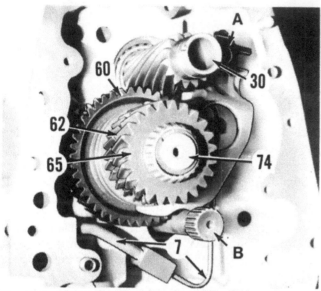

Fig. 136—Front view of transmission with bearing quill removed.

A. Low range I-reverse
   range shift fork
B. Pto drive shaft
7. Lube oil line
30. Low range I-reverse
   drive shaft

60. Gear (Low range I)
62. Gear (reverse)
65. Drive gear (front-wheel
   drive)
74. Extension shaft

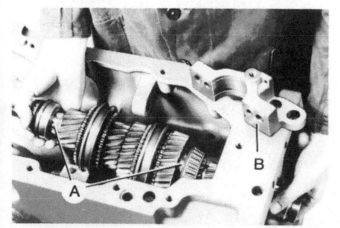

Fig. 137—View showing removal of transmission top shaft assembly.

A. Top shaft assy.

B. Bearing cap

and install snap ring (27) of correct thickness to provide shift sleeve end play of 0-0.1 mm (0-0.004 inch). Snap rings (27) are available in thicknesses of 1.55, 1.65 and 1.75 mm (0.061, 0.065 and 0.069 inch).

Adjust cone point of bevel pinion shaft (44) by installing correct shim pack (47) between bearing cup (46) and transmission case. Measure height of bearing cup (46) and cone (45). Add this measurement to the dimension etched in mm on end of bevel pinion shaft. Subtract the sum of these dimensions from the dimension stamped in identification plate on right side of transmission case. The remainder will be the thickness of shim pack (47) to be installed.

To adjust the bearing preload on bevel pinion shaft (44), install bevel pinion shaft and countershaft cluster gear assembly (44 through 58) in transmission case. Tighten special nut (58 or 58A) to a torque of 140 N·m (100 ft.-lbs.). Engage shift collar (49) in range II. Wrap a string around cluster gear and attach a spring scale to the string to check rolling torque. Rolling drag torque should be 0.75-1.50 N·m (6.5-13 in.-lbs.) with new bearings or 0.40-0.75 N·m (3.5-6.5 in.-lbs.) with used bearings; but when checked with string as described, pull should be 20-40 N (4.5-9 lbs.) with new bearings, 10-20 N (2.2-4.5 lbs.) with used bearings. If necessary, add or remove shims (55) to obtain correct rolling drag torque.

Adjust bearing preload on extension shaft as follows: On models without front-wheel drive, install extension shaft (59 or 59A) on bevel pinion shaft splines. On later models, lugs on extension shaft must engage groove of the later special nut (58A). Install bearing cup (69) so it protrudes about 2 mm (0.078 inch) out front face of bearing quill. Install low range I-reverse shift fork in shift collar (63). Attach bearing quill and tighten retaining cap screws to a torque of 50 N·m (35 ft.-lbs.). Install a thickness of shims (71) between quill and plate (72) sufficient to obtain a measurable end play with cap screws (73) tightened. Measure extension shaft end play. This measurement plus 0.05-0.10 mm (0.002-0.004 inch) is the correct thickness of shims (71) to be removed to obtain correct preload.

On models equipped with front-wheel drive, install extension shaft (74 or 74A) with gears and shift mechanism on bevel pinion shaft splines. On later models, lugs on extension shaft must engage groove of the later special nut (58A). Press bearing cup (76) into bearing quill without shims (78). Attach bearing quill to transmission case and tighten retaining cap screws to a torque of 50 N·m (35 ft.-lbs.). Measure extension shaft end play. This measurement plus 0.05-0.10 mm (0.002-0.004 inch) is the thickness of shim pack (78) to be installed for correct preload.

To adjust preload of bearings on top shaft assembly (Fig. 140), install top shaft assembly and with bearing cups (2 and 32—Fig. 134) in place, but do not install snap ring (1). Install snap ring (33) or retainer plate

(33A). If retainer plate (33A) is used, coat threads of Torx screws with Loctite 242 and tighten to 60 N·m (45 ft.-lbs.) torque. Push assembly, including bearing cup (2), toward front and install the thickest snap ring (1) that can be installed in groove. Install gear (30B) if so equipped, install bearing cup (36) in bearing quill; then install bearing quill. Tighten bearing quill cap screws to a torque of 50 N·m (35 ft.-lbs.). Slide spacer (34) with chamfered side rearward on shaft (30). Install adequate shims (35) to provide measurable shaft end play, with bearing cone (37) and special nut (38) installed. Tighten the special nut (38) to 140 N·m (100 ft.-lbs.) torque and measure shaft end play. This measurement plus 0.05-0.10 mm (0.002-0.004 inch) is the correct thickness of shims (35) to be removed to obtain correct bearing preload. Install bearing cap (B—Fig. 137) and tighten cap screws to 120 N·m (85 ft.-lbs.) torque. Rolling torque of the assembled top shaft should be 1-2 N·m (9-18 in.-lbs.) and can be checked using a torque wrench on nut (38—Fig. 134).

Refer to paragraph 127 and complete the reassembly of tractor.

TRANSMISSION OIL PUMP

The transmission oil pump used is an internal gear, crescent-type pump located in the clutch housing and driven by the pto clutch drive shaft. The pump supplies oil for the Hi-Lo clutch, pto clutch, front-wheel drive clutch and the main hydraulic pump. Oil from the transmission oil pump is also delivered to the oil cooler, to tractor brakes and to the lubricating systems for the Hi-Lo unit and synchronized transmission.

131. A two stage bypass valve (1 through 6—Fig. 141) is located above the filter in transmission housing. If the filter (16) becomes clogged or if oil is too cold to circulate through filter easily, the differential pressure (before the filter and after the filter) will exceed the opening pressure of the bypass valve. At the first stage (about 500 kPa or 70 psi difference), unfiltered oil will be directed into the lube circuits. When the pressure differential exceeds 650 kPa (90 psi), unfiltered oil is still directed to the lubrication circuits, but a larger volume of oil is directed back to the main oil reservoir (transmission case). Be sure shoulder on valve (3) is down when installing. Specifications for spring (5) should be as follows:

Free length . . . . Approx. 68 mm
(2.68 inches)
Pressure when compressed to: . . . . 49 mm/50-60 N
(1.5 inches/11-13 lbs.)
Pressure when compressed to: . . . . 38 mm/80-95 N
(1.93 inches/18-21 lbs.)

If volume of flow is suspected to be low, remove, clean and inspect transmission pump inlet screen (7).

Fig. 139—When installing shift collar, align third tooth of shift collar with thrust block (14) as shown at arrow.

Fig. 140—On later tractors, align arrow mark on shaft (59A or 74A) with groove in special nut (58A). If shaft will not engage groove in nut, turn shaft slightly and re-try.

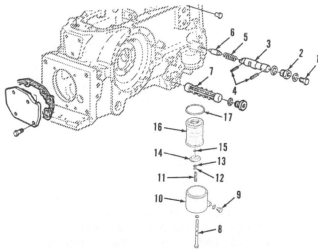

Fig. 141—View showing locations of transmission oil pump screen (7), filter (16) and by-pass valve (3).

1. Plug	
2. Threaded nipple	10. Filter cover
3. By-pass valve	11. Compression spring
4. Spring pins	12. Washer
5. Spring	13. "O" ring
6. Pin	14. Retainer
7. Inlet screen	15. Snap ring
8. Bolt	16. Filter
9. Plug	17. Seal washer

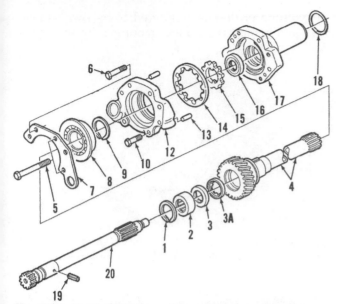

Fig. 142—Exploded view of transmission oil pump assembly typical of all models with synchronized transmission.

1. Snap ring	9. Snap ring
2. Needle bearing	10. Cap screw (2)
3. Oil seal	12. Bearing housing
3A. Cap (late models)	13. Dowel pins (2)
4. Pto clutch drive shaft	14. Internal gear
5. Cap screw (3)	15. Spur gear
6. Cap screw (2)	16. Oil seal
7. Retainer	17. Pump housing & carrier
8. Ball bearing	18. "O" ring

clutch housing and transmission as outlined in paragraph 127. If so equipped, remove Hi-Lo unit, Creeper transmission, front-wheel drive clutch and pto clutch. Remove oil pump suction and pressure lines and retainers. Remove the three oil pump retaining cap screws, then pull oil pump with pto clutch drive shaft out of clutch housing.

Remove cap screws (6—Fig. 142) and separate pto clutch drive shaft (4) from pump assembly. Remove snap ring (9) and press bearing (8) from shaft. Remove snap ring (1), then using a slide hammer puller, remove needle bearing (2), oil seal (3) and cap (3A). Remove cap screws (10) and separate bearing housing (12) from pump housing and carrier (17). Remove spur gear (15) and internal gear (14) from pump housing and carrier. Remove oil seal (16) and "O" ring (18).

Clean and inspect all parts and renew any showing excessive wear or other damage. Bearing housing (12), internal gear (14), spur gear (15) and pump housing and carrier (17) are available only as an assembly.

Reassemble oil pump by reversing disassembly procedure, keeping the following points in mind: Install gears (14 and 15) in pump housing and carrier (17), chamfered side first. Place a straightedge across sealing surface of housing and carrier and measure clearance between gears and straightedge. Clearance should be 0.03-0.05 mm (0.001-0.002 inch). Clean sealing surfaces of pump housing and carrier (17) and bearing housing (12) with Loctite primer, then coat with Loctite 515. Tighten cap screws (6 and 10) to a torque of 55 N·m (40 ft.-lbs.). Fill pump with transmission oil and check pump for free rotation. Coat sealing surface of clutch housing partition with Loctite 515 and coat carrier sleeve with Moly High-Temperature EP grease. Install oil pump assembly and tighten cap screws to a torque of 55 N·m (40 ft.-lbs.).

The balance of reassembly is the reverse of disassembly. Refer to paragraph 127 and rejoin tractor.

If screen is clean, separate tractor between clutch housing and transmission case and check the following:

a. Pump inlet and discharge tube "O" rings and packings for deterioration.
b. Pump body screws for proper torque.
c. Pump body and gears for excessive wear or scoring.

132. R&R AND OVERHAUL. To remove the transmission oil pump, first split tractor between

DIFFERENTIAL AND FINAL DRIVES

The differential is a four pinion type and is equipped with a differential lock.

Final drives incorporate a planetary gear reduction unit at inner end of the housing.

DIFFERENTIAL

133. REMOVE AND REINSTALL. To remove differential, drain transmission and remove final drives as outlined in paragraph 139 or 147 and rockshaft housing as in paragraph 177.

The differential lock must be removed as follows: Remove clamp screw and lever (23—Fig. 143) from

models without cab, or remove bellcrank from end of shaft (20) on models with Sound Gard Body; then remove the square key (21) from all models. Hold fork (16) in place, bump shaft (20) rearward and remove key (19) and plug (17). Remove shaft, fork and shift collar (1).

Remove the transmission oil cup and rear oil lines. Remove the rockshaft control arm and cam from transmission case. Support differential, unbolt both bearing quills (2 and 11), remove quills, then lift out differential. Be sure to keep shims (3) with correct quill to aid in reassembly. Shims are located between quills and transmission case to adjust backlash of the

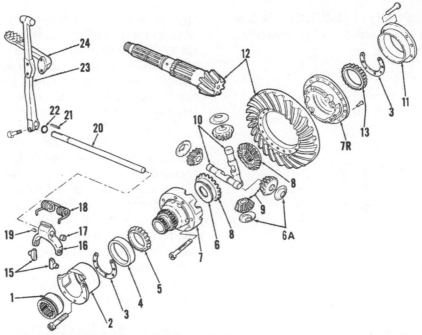

Fig. 143—Exploded view of typical differential and differential lock assembly. Some models do not use thrust washers (6 and 6A). Bevel pinion shaft and ring gear (12) are available only as a matched set for all models.

1. Differential lock collar
2. Left quill
3. Shims
4. Bearing cup
5. Bearing cone
6. Thrust washer
6A. Thrust washer
7. Left housing half
7R. Right housing half
8. Side gears
9. Pinions
10. Pinion shafts
11. Right quill
12. Bevel pinion shaft & ring gear
13. Bearing cone
15. Shoes
16. Fork
17. Plug
18. Spring
19. Woodruff key
20. Shaft
21. Square key
22. "O" ring
23. Lever
24. Pedal

main drive bevel gears and to adjust preload of differential carrier bearings.

When reassembling, first check and adjust mesh position as described in paragraph 136, then check and adjust bearing preload as in paragraph 137 and backlash as in paragraph 138.

134. OVERHAUL. The bevel ring gear, bevel pinion shaft and the right half of the differential housing are available only as a matched set.

To disassemble the removed differential assembly, remove the eight cap screws that attach differential housing halves together. Procedure for removing bearing cup (4—Fig. 143) and bearing cones (5 and 13) will be obvious. Cup for bearing cone (13) is integral with quill (11). If pinions (9) or shafts (10) are worn or damaged, all mating parts should be renewed. Check condition of differential housing bores if side gears (8) are worn. Some models are not equipped with washers (6 and 6A). Inspect differential against the following values:

ID of bevel pinions (9) . . . . 21.82-21.87 mm
(0.859-0.861 in.)
OD of bevel pinion shafts (10) . . . . . 21.70-21.72 mm
(0.854-0.855 in.)
OD of side gear (8) hub . . . . 51.99-52.01 mm
(2.047-2.048 in.)

ID of bores in differential housing . 52.06-52.11 mm
(2.050-2.052 in.)
Thrust washer (6) . . . . 1.95-2.05 mm
(0.077-0.081 in.)
Thrust washer (6A) . . . . 0.95-1.05 mm
(0.037-0.041 in.)

Reassemble by reversing disassembly procedure. Coat shaft (20) with Molykote grease before assembling. Tighten cap screws attaching differential housing halves together, bearing quill to transmission case cap screws and the differential lock operating lever clamp screw all to 50 N·m (35 ft.-lbs.) torque. Final drive to transmission case cap screws should be tightened to 120 N·m (85 ft.-lbs.) torque and the rockshaft housing to transmission case cap screws should be tightened to 120 N·m (85 ft.-lbs.) torque.

MAIN DRIVE BEVEL GEARS

135. ADJUSTMENT. If differential is removed for access to other parts and no defects in the adjustments are noted, the shim pack (53—Fig. 124) and (3—Fig. 143) should be kept intact and reinstalled in their original positions. However, if bevel gears, bearings, quills or transmission case are renewed, the main bevel gears should be checked for mesh (cone point) position, differential carrier bearing preload and gear backlash.

136. MESH (CONE POINT) POSITION. The fore and aft position of the bevel pinion shaft is controlled

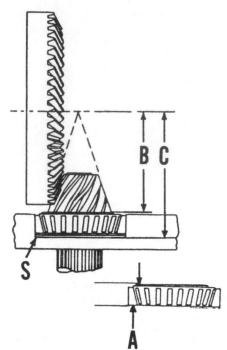

Fig. 143A—Refer to text for setting mesh position of the main drive bevel gears.

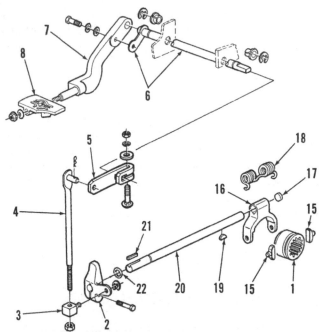

Fig. 144—Exploded view of differential lock linkage used on models equipped with Sound Gard Body.

1. Differential lock collar	15. Shoes
2. Lever	16. Fork
3. Swivel	17. Plug
4. Rod	18. Spring
5. Arm	19. Woodruff key
6. Shaft	20. Shaft
7. Lever	21. Square key
8. Pedal	22. "O" ring

by shims located between pinion shaft rear bearing cup and front wall of differential compartment. Shims for collar shift transmission is shown at (39—Fig. 124) and at (47—Fig. 134) for synchronized transmission. When renewing parts, the shim pack required for correct mesh position of the bevel pinion shaft and ring gear can be determined as follows:

Measure height of the rear pinion bearing cup and cone (A—Fig. 143A). Add 0.10 mm (0.004 inch) to this measurement to compensate for change in height caused by pressing cup into housing. Add this measurement to the dimension etched in mm on end of bevel pinion shaft. Subtract the sum of these dimensions from the dimension stamped on rear top of transmission case or on data plate attached to transmission case. The remainder will be the thickness of shim pack (S) to be installed. Refer to paragraph 123B or paragraph 130 for adjustment of bevel pinion shaft bearing preload.

137. DIFFERENTIAL BEARING ADJUSTMENT. Differential carrier bearings should be adjusted to a preload of 0.08-0.13 mm (0.003-0.005 inch) and adjustment is made as follows: Install differential and bearing quills with original shim packs, then check differential end play using a dial indicator.

NOTE: When making this adjustment, be sure that clearance exists between the main drive bevel ring gear and pinion shaft at all times.

If no differential end play exists, add shims under right bearing quill to obtain not more than 0.05 mm (0.002 inch) end play. If more than 0.05 mm (0.002 inch) end play existed on original check, subtract shims.

On 2150, 2255, 2350 and 2550 models, measure end play of differential, then subtract shims equal to the measured end play plus an additional 0.15-0.25 mm (0.006-0.010 inch) to give the desired bearing preload of 0.15-0.25 mm (0.006-0.010 inch).

On 2155, 2355, 2355N and 2555 models, measure end play of differential, then subtract shims equal to the measured end play plus an additional 0.08-0.13 mm (0.003-0.005 inch) to give the desired bearing preload of 0.08-0.13 mm (0.003-0.005 inch).

Shims are available in thicknesses of 0.08, 0.13 and 0.25 mm (0.003, 0.005 and 0.010 inch).

138. BACKLASH ADJUSTMENT. With differential carrier bearing preload adjusted as in paragraph 137, adjust backlash between bevel pinion shaft and ring gear to 0.3 mm (0.012 inch) by transferring bearing quill shims (3—Fig. 143) from one side to the other as required. Moving shims from right to left will decrease backlash. Do not change total thickness of all shims during backlash adjustment or the previously determined preload adjustment will be changed. Set backlash as close to 0.3 mm (0.012 inch) as possible. Measure backlash at several locations around ring gear. Backlash should not be less than 0.22 mm (0.009 inch), nor more than 0.38 mm (0.015 inch) at any position around ring gear. Improper installation of ring gear is indicated if these limits are exceeded.

FINAL DRIVES

Models 2150, 2155, 2255, 2350, 2355, 2550 W/Collar Shift, 2555 W/Reverser

139. REMOVE AND REINSTALL. To remove a final drive unit, disconnect battery ground cables and drain transmission. Support rear of tractor and remove wheel and tire assembly. Remove rear fenders and roll guard. If right final drive is being removed and tractor has selective control valve, disconnect pressure line, coupler lines and return hose between valve and rockshaft housing and remove control valve. Disconnect brake line from final drive housing. Attach hoist to final drive, remove attaching cap screws and pull final drive assembly from transmission case. If tractor is equipped with Sound Gard Body and only one final drive is to be removed, it may be possible to disconnect body and raise enough to permit removal; however, if both final drives are to be

removed, the body should be removed. Refer to paragraph 186 for removing and installing Sound Gard Body.

Reinstall by reversing removal procedure and tighten attaching cap screws to a torque of 120 N·m (85 ft.-lbs.).

140. OVERHAUL. To overhaul the removed final drive unit, refer to Fig. 145 and proceed as follows:

Remove lock plate (23), cap screw (24) and retainer washer (25), then pull planet carrier assembly (26) from axle. Support outer end of final drive housing (11) so oil seal (2) will clear and press or drive axle out of housing. The axle bearings, bearing cups and oil seals are now available for inspection or renewal. If outer bearing (4) is renewed, heat it to not more than 150° C (300° F), and drive it into place while hot. Be sure that seal spacer is in place before installing bearing cone (4).

NOTE: If axle is flanged, be sure that oil seal is on axle, metal side out, before installing outer bearing.

Bearing cups are pressed in bores until they bottom. Seal cup (3) will be pushed out when outer bearing cup (5) is removed. Be sure to reinstall seal cup after bearing cup is installed. If ring gear and/or

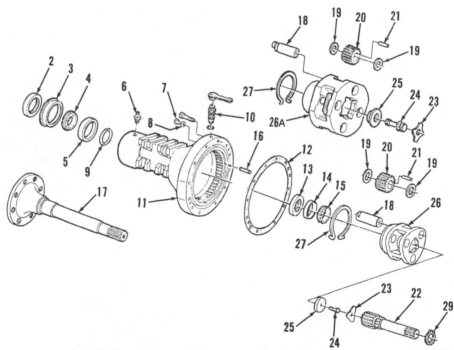

Fig. 145—Exploded view of final drive assembly typical of standard type used on most models. Ring gear is part of axle housing (11).

2. Outer oil seal	9. Snap ring	16. Dowel pin	23. Lock plate
3. Seal cup	10. Brake bleed valve	17. Axle	24. Cap screw
4. Bearing cone	11. Axle housing	18. Planet shaft (3)	25. Washer
5. Bearing cup	12. Gaskets	19. Washers	26. Planet carrier (3 planet)
6. Grease fitting	13. Inner oil seal	20. Planet gear (3)	26A. Planet carrier (4 planet)
7. Plug	14. Bearing cup	21. Bearing needles	27. Snap ring
8. "O" ring	15. Bearing cone	22. Final drive shaft	29. Snap ring

final drive housing is damaged, renew the complete unit.

To remove planet pinions (20), expand snap ring (27), lift it from groove in carrier (26) and pull planet shafts (18) out far enough so snap ring can be removed. Remove planet shafts, planet gears and bearing needles. Check all planetary parts for pitting, scoring or excessive wear and renew parts as required. If any planet gear bearing needles are defective, renew the complete set.

The final drive on 2150, 2155 and 2255 models uses a three planet reduction (26); planetary reduction unit (26A) on other models has four planet gears but is otherwise similar. Each planet gear uses 23 needle rollers.

Before installing final drive, pull final drive shaft (22) and brake disc and inspect. Refer to paragraph 153 for information on brake pressure plate and pressure ring.

Reassemble final drive and adjust axle bearings as follows: Coat bores of planet gears (20) with grease and position rollers (21) in pinions and a thrust washer (19) on each side. Position pinion assemblies (19, 20 and 21) in carrier (26 or 26A) and insert planet shafts (18) only far enough to retain rollers and thrust washers. Install snap ring (27) in slots of planet shafts, then complete insertion of planet shafts and be sure snap ring seats in groove in carrier. Coat inner seal (13) with grease and install axle in housing. Heat inner bearing (15) to not more than 150° C (300° F), and install on inner end of axle. Place carrier assembly on axle, install retaining washer (25) and cap screw (24). Tighten cap screw until bearing is pulled into place and a small amount of axle end play remains. Check the amount of torque required to turn the axle with the existing axle end play. Then, tighten the cap screw to increase the rolling torque an additional 11-17 N·m (98-150 in.-lbs.) for 2550 with collar shift transmission, 2150, 2350 and 2255 models; or 8-12.5 N·m (72-108 in.-lbs.) for 2555 models with reverser and all 2155 models. Install lock plate (23).

Fill axle outer bearing opening with multipurpose grease and install oil seal (2) with metal side out.

Models 2355N, 2550 With Synchronized Transmission, 2555

147. REMOVE AND REINSTALL. To remove a final drive unit, disconnect battery ground cables and drain transmission. Support rear of tractor and remove wheel and tire assembly. Remove rear fenders and roll guard. If right final drive is being removed and tractor has selective control valve, disconnect pressure line, coupler lines and return hose between valve and rockshaft housing and remove control valve. Disconnect brake line from final drive housing. Attach hoist to final drive, remove attaching cap screws and pull final drive assembly from transmission case. If tractor is equipped with Sound Gard Body and only one final drive is to be removed, it may be possible to disconnect body and raise enough to permit removal; however, if both final drives are to be removed, the body should be removed. Refer to paragraph 186 for removing and installing Sound Gard Body.

Reinstall by reversing removal procedure and tighten attaching cap screws to a torque of 120 N·m (85 ft.-lbs.).

148. OVERHAUL. To overhaul the removed final drive unit, first remove brake housing with ring gear (10—Fig. 148) from final drive housing (19). Remove lock plate (25), unscrew special twelve-point cap screw (26), then pull planetary carrier assembly from axle. Drive rear axle shaft out of final drive housing and remove bearing cone (5). Remove snap ring (4), bearing cone (22) and oil seal (1). To remove planet gears (33), expand snap ring (28), lift it from groove in carrier (29) and pull planet shafts (34) out far enough so snap ring can be removed. Remove planet shafts, planet gears and bearing needles. Check all

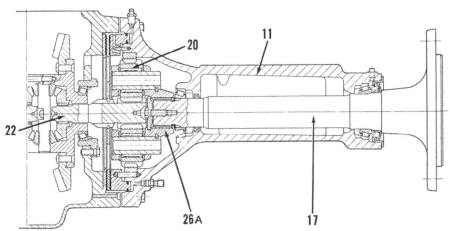

Fig. 145A—Cross section of final drive shown in Fig. 145.

planetary parts for pitting, scoring or excessive wear and renew parts as required. If any planet gear bearing needles are defective, renew the complete set.

Reassemble final drive and adjust axle bearings as follows: Coat bores of planet gears with grease and place needles in gears. Install spacer (32) between the two rows of bearing needles and position a thrust washer at each side. Place planet gears in carrier and insert planet shafts (34) only far enough to retain bearing needles and thrust washers. Install snap ring (28) in slots of planet shafts, then complete insertion

of shafts and be sure snap ring seats in groove in carrier.

Heat bearing cone (22) to 150° C (300° F) and drive on rear axle shaft. Coat inner seal (7) with grease and install axle in housing. Heat bearing cone (5) to not more than 150° C (300° F), and install on inner end of axle shaft. Place carrier assembly on axle, install retaining washer (27) and cap screw (26). Tighten cap screw until bearing is pulled into place and a small amount of axle end play remains. Check the amount of torque required to turn the axle with the existing axle end play. Then, tighten the cap screw to increase

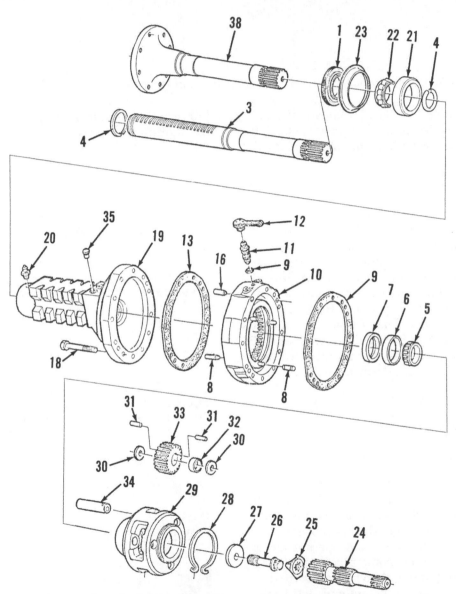

Fig. 148—Exploded view of heavy duty final drive assembly used on some models.

1. Oil seal	10. Brake housing with ring gear	21. Bearing cup	29. Planet carrier
3. Axle shaft	11. Brake bleed valve	22. Bearing cone	30. Washers
4. Snap ring	12. Cap	23. Seal cup	31. Bearing needles
5. Bearing cone	13. Gasket	24. Final drive shaft	32. Spacer
6. Bearing cup	16. Dowel pin	25. Lock plate	33. Planet gear (3)
7. Oil seal	18. Cap screw	26. Special cap screw	34. Planet shaft (3)
8. Dowel pin	19. Final drive housing	27. Washer	35. Plug
9. Gasket	20. Grease fitting	28. Snap ring	38. Flanged axle shaft

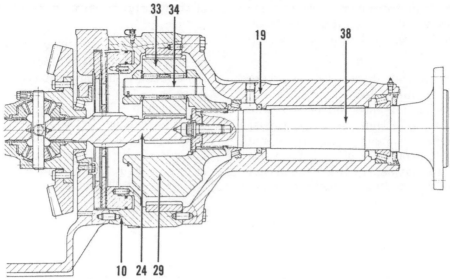

Fig. 148A—Cross section of heavy duty final drive and axle shown in Fig. 148.

the rolling torque an additional 10-14.5 N·m (90-126 in.-lbs.). Install lock plate (25). Use new gaskets (9 and 13) and reinstall brake housing and final drive.

Lubricate outer axle bearing (22) with 6-8 strokes of EP multipurpose grease.

BRAKES

The brakes are hydraulically actuated and utilize a wet-type disc controlled by a brake operating valve located on right side of clutch housing on models not equipped with Sound Gard Body or on right side of dash panel support on models with Sound Gard Body. Brake discs are splined to the final drive shafts and the brake pressure ring is fitted in inner end of final drive housing. Except for a pedal adjustment, no other brake adjustments are required.

BLEED AND ADJUST

All Models

149. BLEEDING. Brakes must be bled when pedals feel spongy, pedals bottom, or after disconnecting or disassembling any portion of the brake system.

To bleed brakes, start engine and operate for at least two minutes at 2100 rpm, turning steering wheel from left to right and back to left to ensure that air is removed from steering system and that brake control valve reservoir is filled. Attach a clear plastic bleed hose to brake bleed screw (10—Fig. 145) or (11—Fig. 148) located on top side of final drive housing and place opposite end in filler hole of rockshaft housing. Loosen bleed screw about ¾ turn. Depress brake pedal being bled. Before pedal reaches the end of its travel, close bleeder screw and let pedal return.

Continue sequence until oil in bleed hose is free of air bubbles.

NOTE: If engine is not running, engine must be started and brake valve reservoir refilled after each 15 strokes of the pedal.

Remove bleed hose and repeat operation on the other brake.

150. PEDAL ADJUSTMENT. Before making this adjustment, bleed brakes as in paragraph 149. For access to brake control valve on models with Sound

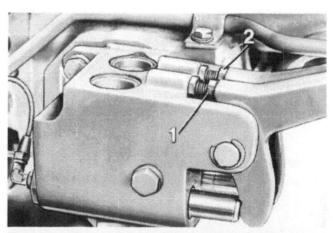

Fig. 149—View of brake pedal adjusting screws on models not equipped with Sound Gard Body. Adjustment is similar on models with Sound Gard Body.

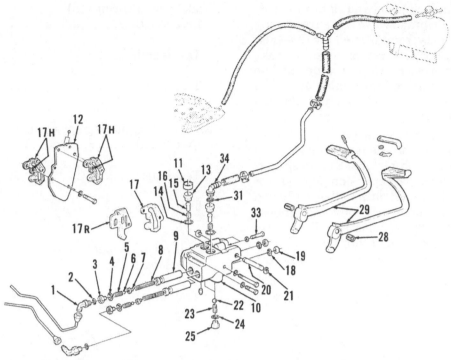

Fig. 150—Exploded view of brake control (master cylinder) valve typical of models without Sound Gard Body.

1. Fitting
2. "O" ring
3. Check valve seat
4. "O" ring
5. Spring
6. Steel ball
7. Retainer
8. Spring
9. Piston
10. Valve housing
11. Expansion plug
12. Intermediate plate (models w/high platform)
13. Check valve seat
14. "O" ring
15. Valve
16. Spring
17. Standard gasket
17H. Holes for high platform gasket
17R. Gasket with Direction Reverser
18. "O" ring
19. Oil seal
20. Pedal shaft
21. Retaining ring
22. Steel ball
23. Spring
24. "O" ring
25. Plug
28. Pedal bushing
29. Pedals
33. Cap screw

Gard Body, raise platform mat, then remove hand brake lever cover, hand brake lever, rear dash panel and right side dash panel.

On all models, adjust stop screws (1 and 2—Fig. 149) so that brake pistons are fully extended from housing and brake pedal arms are just touching the pistons. Then, turn both stop screws out an additional ⅓ turn. If pedals are not the same height, turn the stop screw on highest pedal out ⅙ turn.

BRAKE TEST

All Models

151. PEDAL LEAK-DOWN. With a 270 N (60 lbs.) pressure applied continuously to each pedal for one minute, the pedal leak-down should not exceed 25 mm (1 inch). Excessive brake pedal leak-down can be caused by air in the brake system, faulty brake control valve pistons and/or "O" rings, faulty brake pressure ring seals, or faulty brake control valve equalizing valves or reservoir check valves. Faulty brake control valve pistons or "O" rings will be indicated by external leakage around the pistons.

Faulty pressure ring seals, or brake control valve, can be determined as follows: Isolate brake from brake valve by plugging brake line. If leak-down

stops, pressure ring seals are defective. If leak-down continues, brake control valve is faulty and can be checked further by depressing brake pedals individually, then simultaneously. If leak-down occurs in both cases, a defective reservoir check valve is indicated. If leak-down occurs during individual pedal operation, but not on simultaneous pedal operation, a faulty equalizer valve is indicated.

Refer to appropriate following paragraphs for brake control valve and brake pressure ring overhaul procedures.

R&R AND OVERHAUL

All Models

152. BRAKE CONTROL VALVE. To remove the brake control valve from models not equipped with Sound Gard Body, thoroughly clean valve and surrounding area. Disconnect brake lines from rear of control valve and cap or plug openings. Remove retaining ring (21—Fig. 150) and remove pedal shaft (20). Remove brake pedals, then unbolt and remove valve assembly.

On models equipped with Sound Gard Body, raise platform mat, then remove hand brake lever cover, hand brake lever, rear dash panel and right side dash panel. Thoroughly clean brake valve and surrounding area. Disconnect oil lines from inlet elbow (30—Fig. 151) and outlet elbow (34), then disconnect brake lines from elbow fittings (1). Cap or plug all line openings. Unbolt and remove brake valve assembly. Remove retaining ring (21), withdraw shaft (20) and remove brake pedals (29).

On all models, remove fittings (1—Figs. 150 or 151), springs (5) and balls (6). Remove seats (3) and ball retainers (7), then push pistons (9) and springs (8) out of valve bores. Remove expansion plugs (11) or connectors (32) with inlet and outlet elbows (30 and 34), then remove reservoir check valve assemblies (items 13, 14, 15 and 16). Remove equalizer valve assemblies (items 22, 23, 24 and 25). "O" rings (18) and oil seals (19) can be removed from piston bores.

Clean and inspect all parts for wear, scoring, cracks or other damage. Renew as required. Check springs as follows:

Brake piston springs (8)—
  Free length .... 190 mm
                                     (7.5 in.)
  Test length .... 145 mm
                                       (5.75 in.)
  Test load .... 90 N
                                       (20 lbs.)

Return oil regulating springs (5)—
  Free length .... 11 mm
                                       (0.43 in.)
  Test length .... 6 mm
                                       (0.24 in.)
  Test load .... 0.8-0.9 N
                                   (2.7-3.3 oz.)

Check valve springs (16)—
  Free length .... 23 mm
                                       (0.90 in.)
  Test length .... 8 mm
                                       (0.32 in.)

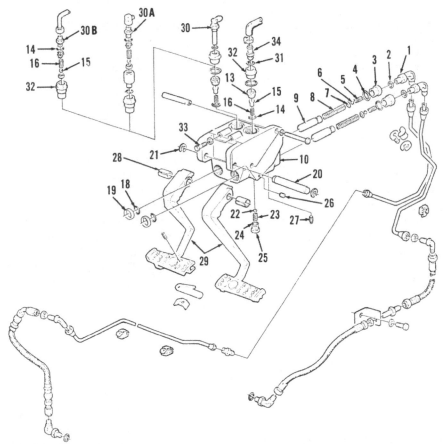

Fig. 151—Exploded view of brake control (master cylinder) valve typical of models with Sound Gard Body. Parts shown with adapter (30A) are used with early models with return oil filter; parts shown with adapter (30B) are used on later models with return oil filter.

1. Elbow fitting	8. Spring	18. "O" ring
2. "O" ring	9. Piston	19. Oil seal
3. Check valve seat	10. Valve housing	20. Brake pedal shaft
4. "O" ring	13. Check valve seat	21. Retaining rings
5. Spring	14. "O" ring	22. Steel ball
6. Steel ball	15. Check valve	23. Spring
7. Retainer	16. Spring	24. "O" ring

25. Plug	30A. Early type adapter w/filter
26. Set screw	30B. Later type adapter w/filter
27. Plug	31. "O" ring
28. Pedal bushing	32. Connector
29. Brake pedals	33. Cap screw
30. Oil inlet elbow w/o filter	34. Oil return connector

Test load . . . . .0.4-0.6 N
(1.4-2.0 oz.)

Equalizer valve springs (23)—
Free length . . . . .20 mm
(0.80 in.)
Test length . . . . .7 mm
(0.28 in.)
Test load . . . . .0.6-0.8 N
(2.0-2.7 oz.)

Renew housing if seats for equalizer balls (22) are damaged. Oil seals (19) are installed with lips toward outside. Pay particular attention to area of reservoir check valves (15) where contact is made with brake pistons and renew valves if any doubt exists as to its condition. Renew bushings and/or pedal shaft (20) if clearance is excessive. Lubricate all parts when reassembling.

Fig. 152—Brake disc is splined to final drive shaft. Model shown is typical.

Fig. 153—View of pressure ring removed from brake housing. Model shown is typical but differences will be noted.

1. Pressure ring
2. "O" rings

Use new gasket (17—Fig. 150) when installing brake valve on tractor not equipped with Sound Gard Body. Gasket for models with direction reverser is shown at (17R) and gaskets used with intermediate plate (12) have holes shown at (17H). Reinstall brake valve by reversing removal procedure. Bleed brakes as in paragraph 149 and adjust pedals as in paragraph 150.

153. BRAKE PRESSURE RING, PLATE AND DISC. To remove pressure ring, plate and disc, remove final drive as outlined in paragraphs 139 or 147. Pull final drive shaft from differential and remove brake disc from final drive shaft. See Fig. 152. Lift brake pressure plate from dowels in transmission case. Remove pressure ring (Fig. 153) from brake housing.

Inspect brake disc for worn or damaged facings or damaged splines and renew as necessary. Inspect pressure plate for scoring, checking or other damage and renew as required.

To reassemble, install new seals on pressure ring, lubricate and install in brake housing over dowel pins. Make sure that neither seal is cut or rolled during installation. Place pressure plate over dowels in transmission case. Install brake disc on final drive shaft so thickest facing is next to transmission case and insert shaft into differential. Install final drive housing.

Bleed brakes as outlined in paragraph 149.

PARKING BRAKE

R&R AND OVERHAUL

154. On models with Sound Gard Body, remove the body as outlined in paragraph 186. On all tractors, drain transmission and remove rockshaft housing assembly as in paragraph 177 and the left final drive assembly as outlined in paragraphs 139 or 147. If the brake drum (9—Figs. 154 or 154A) is to be removed, refer to paragraph 133 and remove the differential.

On models without body, remove hand brake lever (1—Fig. 154) and Woodruff key from shaft (22) or lever (21). On models with Sound Gard Body, refer to paragraph 186 to remove the body, then remove lever (27—Fig. 154A) and Woodruff key from shaft (22). On all models, remove support screw (11—Fig. 154 or Fig. 154A) from bottom of transmission case and remove brake adjusting screw (15) from threaded end of brake band (12). Use a ⅜ inch-16 UNC cap screw to pull upper anchor pin (19) and lower anchor pin (20) from bores. Remove pin and detach link (18) from shaft (22), then remove shaft (22). Pull anchor (17) up far enough to remove pin (13), then pull anchor (17) and associated parts from transmission case. Rotate brake band enough to remove spring (14), unhook spring (14), then remove band (12).

To remove brake drum and hub (9—Figs. 154 or 154A), first remove differential as outlined in paragraph 133.

Clean and inspect all parts for excessive wear or other damage and renew as required. If new linings are riveted to band, heads should be toward outside. Reinstall hand brake by reversing the removal procedure and coat brake shaft (22) and bore in case with Moly High Temperature EP grease. If new hand brake linings (12) have been installed, adjust band support screw (11) by loosening two turns and tightening lock nut. Re-adjust support screw after about 100 hours to the standard setting.

Adjust brake as follows: Apply hand brake lever so that latch (5) engages first notch of quadrant (6). Tighten brake band adjusting screw (15) by hand

using a screwdriver. Tighten band support screw (11) at bottom of transmission case by hand using a screwdriver, then back off ½ turn and secure with locknut. Adjust brake band adjusting screw (15) until hand brake lever latch (5) will engage in third or fourth notch in quadrant when lever is pulled with a force of about 110 N (25 lbs.).

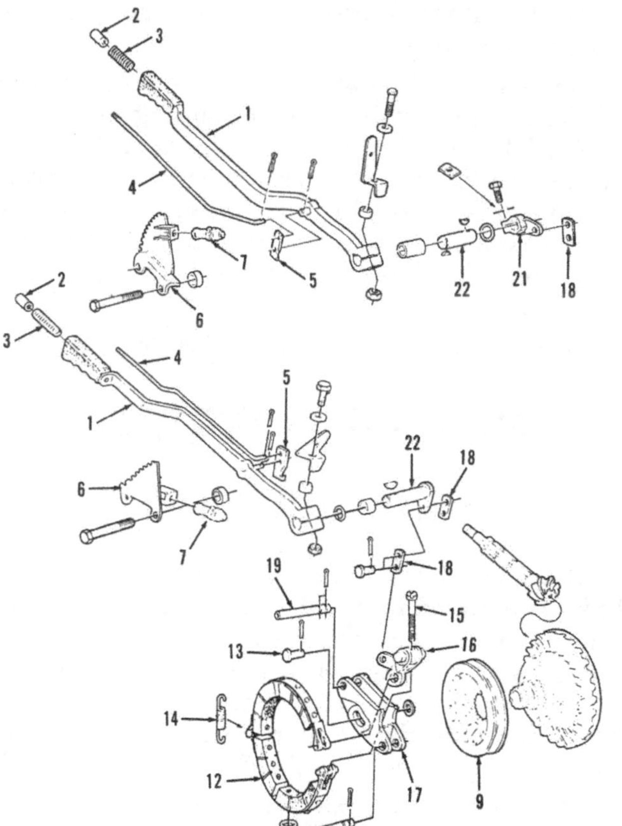

POWER TAKE-OFF (INDEPENDENT)

OPERATION

155. The Independent Power Take-Off may provide only one speed (540 rpm) or dual speeds (either 540 rpm or 1000 rpm) operation. On dual speed pto, the speed will depend upon which stub shaft (20 or 21—Fig. 158B) is installed. On all models, engagement of the multiple disc hydraulic clutch or the pto brake is accomplished by directing oil with the pto control valve (Fig. 157). The pto control valve is designed so there is no overlap between clutch and brake circuits. When control lever is moved to the engaged position, pressure fluid flows to the pto clutch and at the same time enters the area behind the valve. This prevents rapid opening of the valve and consequent rough clutch engagement. At the same time pressure is

Fig. 154—Exploded view of hand brake assembly typical of models without Sound Gard Body.

1.	Brake lever	13.	Pin
2.	Release button	14.	Spring
3.	Spring	15.	Adjusting screw
4.	Release rod	16.	Brake band arm
5.	Latch	17.	Anchor
6.	Quadrant	18.	Links
7.	Brake light switch	19.	Upper anchor arm pin
9.	Brake drum	20.	Lower anchor arm pin
11.	Band support screw	21.	Lever
12.	Brake band	22.	Brake shaft

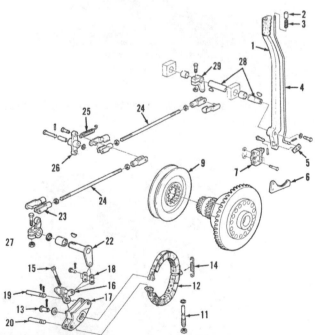

Fig. 154A—Exploded view of hand brake assembly typical of models with Sound Gard Body. Refer to Fig. 154 for legend except for the following.

23.	Yoke		
24.	Brake rod	27.	Lever
25.	Spring	28.	Shaft
26.	Pivot arm	29.	Arm

ported to the clutch, pto brake pressure is released and the brake piston passage is opened to the sump.

When control lever is moved to the disengaged position, clutch passage is opened to the sump and pressure is applied to a piston located in countershaft bearing support that applies a disc-type brake to pto clutch shaft.

NOTE: When there is a drop in hydraulic pressure or when engine is shut off, the hydraulic shift lever arm detent releases and control lever automatically shifts to disengaged (brake) position.

PRESSURE TEST

155A. On tractors without Sound Gard Body, remove shift lever shield from transmission shift cover. On models equipped with Sound Gard Body, remove floor mat and floor plate. On all models, to check pto brake pressure, remove plug at (E—Fig. 156) and install test gage. To check pto clutch pressure, remove plug (A) and install test gage. With pto control lever in neutral position, start engine and operate at 1500 rpm. Move lever to brake or clutch position. Gage should read 1000 kPa (145 psi) in either position.

With gages installed in ports (A and E) and engine operating at 1500 rpm, observe both gages. Slowly move control lever from engaged to disengaged position. One gage must read zero before other gage shows any pressure. No pressure overlap should be experienced when shifting slowly between engaged and disengaged positions.

R&R AND OVERHAUL

156. CONTROL VALVE. The control valve is located in the transmission shift cover. On models without Sound Gard Body, remove shift lever shield from transmission shift cover. On models equipped with Sound Gard Body, remove floor mat and floor plate. On all models, disconnect pto shift linkage from control lever shaft (16 or 18—Fig. 157). Remove cover (20) with shift arm detents (25 and 30) and control lever shaft. Remove transmission shift cover assembly (32). Remove all parts of detents (25 and 30) from cover (20). Drive plug (33) out of transmission shift cover, then remove pto control valve assembly (44). Carefully separate valve spool parts, keeping shims (35) in their respective locations.

Check spool (39) and shift cover bore for damaged lands. Check tension of springs (36, 40 and 42) as follows:

Spring (36)—
Free length . . . . .48 mm
(1.91 in.)
Test length . . . . .34 mm
(1.33 in.)

Test load . . . . .90-110 N
(20-25 lbs.)

Spring (40)—
Free length . . . . 22 mm
(0.875 in.)
Test length . . . . 16 mm
(0.625 in.)
Test load . . . . .55-65 N
(12-15 lbs.)

Spring (42)—
Free length . . . . 21 mm
(0.810 in.)
Test length . . . . 14 mm
(0.540 in.)
Test load . . . . .200-240 N
(45-55 lbs.)

When reassembling, determine correct shim packs (35) as follows: Install washer (37), spring (36) and one shim (35) and snap ring (34) on special pin (38). Place special pin (snap ring end first) on a spring scale. Press down on washer (37) with your fingers. The washer should break free of head on special pin at 30-34 N (6.5-7.0 lbs.). If necessary, add shims until correct spring preload has been obtained.

NOTE: There must always be at least one shim installed.

Install four shims (35—Fig. 157A), spring (40), sleeve (41), spring (42), three shims (35), actuator (43) and snap ring (34) on spool (39). Place large end of spool on spring scale and press with fingers on actuator. At a load of 57-61 N (12-13 lbs.) actuator should

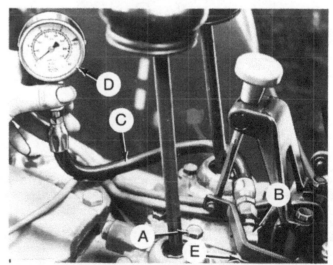

Fig. 156—Use a 2070 kPa (300 psi) test gage to check pto clutch and brake pressure.

A. Pto clutch test port
B. JDG-84 elbow
C. Test hose
D. Test gage
E. Brake test port

break free from snap ring. If not, add or remove shims until correct spring preload has been obtained. When actuator just breaks free from snap ring, clearance between sleeve (41) and shoulder of spool (39) should be 1.5-2.0 mm (0.06-0.08 inch) as shown at (a).

Use all new "O" rings and reassemble by reversing the disassembly procedure.

157. PTO CLUTCH, BRAKE AND INPUT GEARS.
Separate tractor between clutch housing and transmission case as outlined in paragraphs 121 or 127. Remove oil lines (7 and 8—Fig. 158), pull connector (11) out to the front and remove suction line (13). If so equipped, remove Hi-Lo shift unit, Creeper transmission or front-wheel drive clutch. Remove drive gear to clutch drum Allen screws. Use two of the screws in jack screw holes and remove drive gear. Remove snap ring (1) and pull clutch drum (2) from

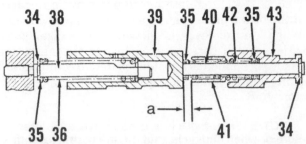

Fig. 157A—Refer to text for adjustment of actuator and spool. Refer to Fig.157 for legend.

pto clutch shaft (10). Remove connector (6) and oil inlet lines (4 and 5). Remove cap screws (5—Fig. 158A) and lift out oil manifold (6) with pto brake and pto clutch shaft (32). Remove seal rings (15) and place clutch drum (17) in a protected vise with clutch side

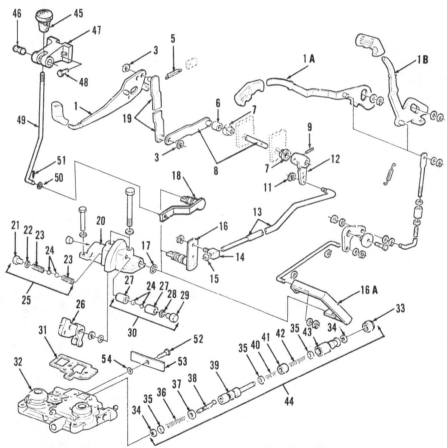

Fig. 157—Exploded view of pto control valve and linkage. Items (1 through 16) are used on tractors with Sound Gard Body. Items (18 and 45 through 54) are used on tractors without Sound Gard Body.

1. Control lever	13. Rod	23. Springs	34. Snap ring
1A. Control lever	14. Swivel	24. Detent balls	35. Shims	45. Knob
1B. Control lever	15. Nut	25. Mechanical detent	36. Spring	46. Pin
3. Snap ring	16. Control lever shaft	26. Control arm	37. Washer	47. Bracket
5. Spring	16A. Control lever shaft	27. Pistons	38. Special pin	48. Cap screw
6. Bushing	17. "O" rings	28. "O" ring	39. Spool	49. Rod
7. Bushing	18. Control lever shaft	29. Plug	40. Spring	50. Washer
8. Lever shaft	19. Link	30. Hydraulic detent	41. Sleeve	51. Cotter pin
9. Spring pin	20. Cover	31. Gasket	42. Spring	52. Countersunk screw
11. Snap ring	21. Plug	32. Shift cover	43. Actuator	53. Retaining plate
12. Lever	22. "O" ring	33. Plug	44. Control valve spool assy.	54. Washer

Fig. 158—Installed pto clutch and oil lines typical of all models.

1. Snap ring	8. Transmission oil filter line
2. Clutch drum	9. Oil pump pressure line
3. Drive gear	10. Pto clutch shaft
4. Pto brake line	11. Oil pump suction connector
5. Pto clutch line	12. Oil pressure line
6. Pto brake inlet connector	(front-wheel drive)
7. Oil cooler line	13. Oil pump suction line

up. Remove snap ring (23) and lift out clutch hub (30), back plate (24), clutch plates (26) and clutch discs (25). Use a JDT-24A spring compressor or equivalent, compress spring (29) and remove snap ring (27). Remove retainer (28) and spring (29). Remove clutch piston (20) and seals (19 and 21). Remove snap ring (2) and withdraw pto clutch shaft (32) with ball bearing (3) from oil manifold (6). Remove snap ring (14), back plate (12), brake disc (13), brake plate (12), then remove brake piston (11) and seal rings (9 and 10).

Clean and inspect all parts and renew any showing excessive wear or other damage. Check spring (29) against the following specifications: Free length should be 98 mm (3.88 inches). Spring tension at a length of 46 mm (1.80 inches) should be 530-650 N (117-143 lbs.).

Use all new seal rings and reassemble by reversing the disassembly procedure, keeping the following points in mind. Tighten oil manifold to clutch housing cap screws (5) to 50 N·m (35 ft.-lbs.) torque. Install spring retainer (28) so that vanes are away from piston (20). Use Fig. 158A as a guide and install correct number of clutch discs (25) and plates (26). Eight plates and disks are used with dual speed pto, ten of each are installed on 540 rpm pto.

Reinstall by reversing removal procedure.

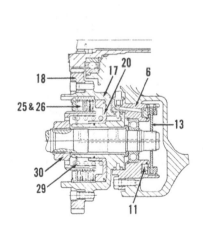

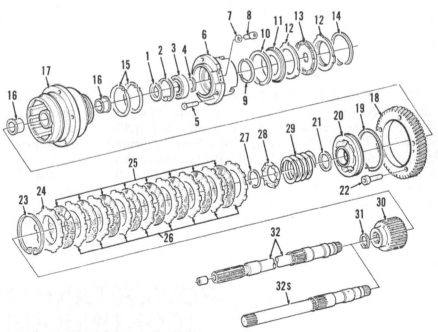

Fig. 158A—Exploded view and cross section of pto clutch and brake assembly.

1. Spacer	9. Seal ring	17. Clutch drum	24. Back plate	31. Snap ring
2. Snap ring	10. Seal ring	18. Drive gear	25. Clutch discs	32. Pto clutch shaft
3. Ball bearing	11. Piston	19. Seal ring	26. Clutch plates	(Collar Shift
4. Snap ring	12. Brake plate	20. Piston	27. Snap ring	transmission)
5. Cap screw (3)	13. Brake disc	21. Seal ring	28. Spring retainer	32S. Pto clutch shaft
6. Oil manifold	14. Snap ring	22. Allen screw	29. Coil spring	(Synchronized
7. Washer	15. Seal rings	23. Snap ring	30. Clutch hub	transmission)
8. Allen screw	16. Bushings			

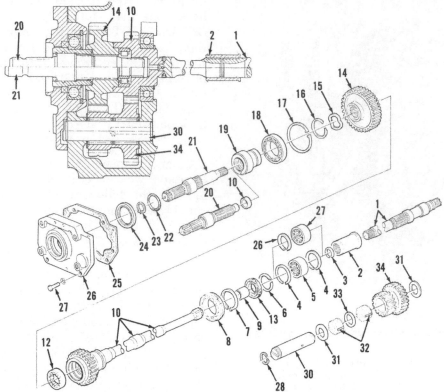

Fig. 158B—Exploded view and cross section of dual speed pto rear shafts and gears.

1. Pto clutch shaft	9. Bushing	18. Ball bearing	26. Bearing quill
2. Sleeve coupler	10. Pto shaft & gear	19. Pilot	27. Cap screw
3. Snap ring	12. Roller bearing	20. 540 rpm stub shaft	28. Snap ring
4. Snap rings	13. Oil seal	21. 1000 rpm stub shaft	30. Countershaft
5. Needle bearing	14. Gear	22. "O" ring	31. Thrust washer
6. Snap ring	15. Spring washer	23. Snap ring	32. Bearing needles
7. Snap ring	16. Snap ring	24. Oil seal	33. Spacer
8. Ball bearing	17. Snap ring	25. Gasket	34. Reduction gear

157A. OUTPUT SHAFT AND GEARS. To remove the output shaft and gears, first drain transmission case, then unbolt and remove bearing quill assembly (16 through 26—Fig. 158B). Remove spring washer (15) and gear (14), then withdraw pto shaft and gear (10). Remove snap ring (28). Pull shaft (30) out of transmission case and remove thrust washers (31) and reduction gear (34) with bearing needles (32) and spacer (33).

Complete disassembly of removed components. To remove sleeve coupler (2) and needle bearing (5), tractor must be split between clutch housing and transmission case.

Clean and inspect all parts and renew any showing excessive wear or other damage. Use all new gasket, oil seal and "O" rings and, using Fig. 158B as a guide, reinstall by reversing the removal procedure.

POWER TAKE-OFF
(CONTINUOUS)

Tractors may be equipped with 540 rpm continuous running pto with either collar shift or synchronized transmission. A mid-pto is also available on some models; however, the mid-pto operates only at 1000 rpm. Shifter couplings are provided to engage and disengage rear pto and mid-pto shafts.

When service is required on the pto system, the following should be taken into consideration. Work involving rear pto shaft can be accomplished by working from rear of tractor. Work involving the pto driven gears, power shaft clutch shaft, mid-pto shaft or mid-pto shifter assembly will require that the clutch hous-

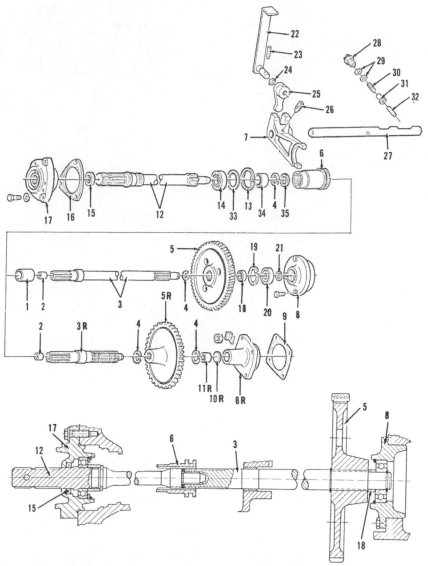

Fig. 159—Exploded and cross section views of continuous running pto used on some models with collar shift transmission. Parts identified with "R" suffix are used with direction reverser.

1. Bushing	9. Gasket	17. Bearing quill	25. Internal shift lever
2. Bushing	10R. Thrust washer	18. Spacer	26. Set screw
3. Drive (front) shaft	11R. Needle bearing	19. Snap ring	27. Shift rod
4. Snap ring	12. Power (rear) shaft	20. Ball bearing	28. Neutral start switch
5. Driven gear	13. Snap ring	21. Snap ring	29. Washers
6. Shift collar	14. Ball bearing	22. Shift lever	30. Spring
7. Shift fork	15. Oil seal	23. Spring pin	31. Guide
8. Cover	16. Gasket	24. "O" ring	32. Pin

ing be separated from the transmission case. Work involving the rear pto shaft shifter assembly will involve removing the rockshaft housing, high and low range shifter shafts and countershaft assembly in addition to separating the clutch housing from transmission case.

REAR PTO SHAFT

158. R&R AND OVERHAUL. To remove the rear pto shaft (12—Fig. 159, Fig. 159A or Fig. 159B), drain transmission and remove pto shield and shaft guard,

if so equipped. Place rear pto shaft control lever in "OFF" position, then remove cap screws from pto shaft bearing quill (17) and pull shaft and quill from transmission case. Be careful not to pull shift collar from front drive shaft. Bearing (14) can be renewed after removing snap rings (4 and 13). Press new oil seal (15) in quill, with lips toward front, until it bottoms. For service on remainder of pto assembly, split tractor as outlined in paragraphs 121 or 127.

Reassemble by reversing removal procedure and mate splines of rear shaft with splines of shift collar as shaft is installed.

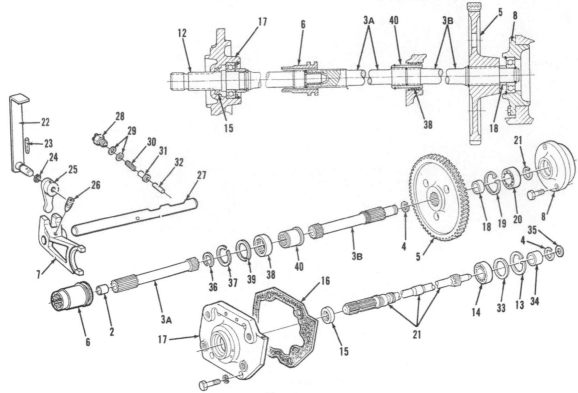

Fig. 159A—Exploded and cross section views of continuous running pto used on some models with synchronized transmission. Refer to Fig. 159 for legend except the following.

3A. Intermediate shaft
3B. Drive (front) shaft
33. Washer
34. Spacer bushing
35. Thrust washer
36. Snap ring
37. Snap ring
38. Needle bearing
39. Snap ring
40. Connecting sleeve

DRIVEN GEARS, FRONT DRIVE SHAFT AND MID-PTO SHAFT

159. R&R AND OVERHAUL. To service the pto driven gears, front power shaft or mid-pto front stub shaft, it is first necessary to split the tractor as outlined in paragraph 121 or 127.

The pto clutch power shaft and engine clutch shaft can also be removed at this time by withdrawing them rearward out of clutch housing. Pto drive gear can be pressed from power shaft if necessary.

Prior to any disassembly, install rear pto shaft, if removed, and place rear pto shaft shifter in "ON" position. This will slide the shift collar on rear pto shaft and prevent it from falling to the bottom of the transmission case. **If shift collar comes off, it will be necessary to remove the rockshaft housing to retrieve it.**

On models with only 540 rpm pto, the power shaft (3—Fig. 159 or 3B—Fig. 159A) and associated parts can be removed after removing the screws attaching front quill (8). Screws are accessible through holes in gear (5). Remove snap ring (21) to withdraw shaft from bearing (20).

On models with 540 rpm rear pto and 1000 rpm mid-pto, remove spring (46—Fig. 159B), ball (47) and driven gears (5 and 45). Pull drive shaft (41) from bore and inspect thrust washer (43). Also inspect bushings (2, 42 and 44), needle bearings (49 and 53) and associated bearing surfaces. Disassembly of the front drive parts (47 through 66) will be evident. Install oil seal (55) with metal side forward. Use a seal protector and caution when inserting shaft (51) through seal (55) to prevent damage to seal lips.

REAR PTO SHAFT SHIFTER

160. R&R AND OVERHAUL. To remove the rear pto shifter assembly, remove rockshaft housing as outlined in paragraph 177 and the transmission countershaft as outlined in paragraph 123 or 130. Remove shifter shaft lock (28 through 32—Fig. 159, Fig. 159A or Fig. 159B), located forward and slightly below pto shift lever. Drive pin (23) from lever (25) and shaft. Remove set screw (26), slide shaft (27) from fork (7) and withdraw parts from housing.

Clean and inspect all parts and renew as necessary. Reinstall shifter mechanism by reversing removal procedure.

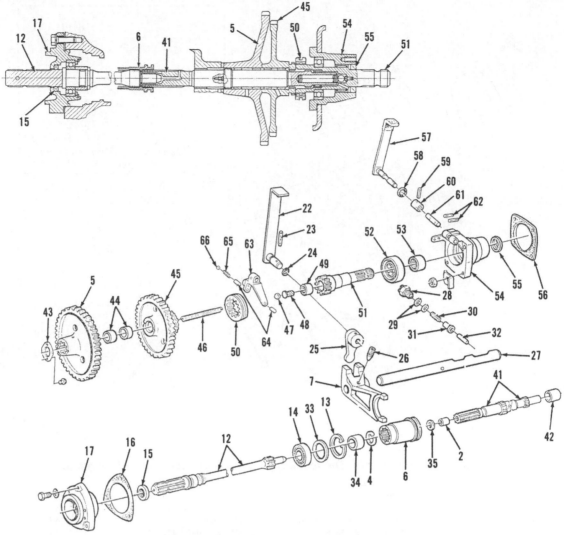

Fig. 159B—Exploded and cross section views of continuous running pto used on some models which provides 540 rpm rear pto and 1000 rpm mid pto. Refer to Fig. 159 for legend except the following.

41. Drive shaft for rear pto	48. Pin	55. Oil seal	61. Shaft
42. Bushing	49. Needle bearing	56. Gasket	62. Spring pins
43. Thrust plate	50. Shift collar	57. Engaging lever	63. Internal shift lever
44. Bushings (4 used)	51. Mid pto output shaft	58. "O" ring	64. Pin
45. Gear for 1000 rpm	52. Ball bearing	59. Spring pin	65. Detent spring
46. Spring	53. Needle bearing	60. Coupler	66. Detent ball
47. Ball	54. Front bearing quill		

FRONT POWER TAKE-OFF

Tractors may be equipped with a pto located at the front of the tractor that is driven by a shaft running through the hydraulic pump. The front pto operates at 1000 rpm and is engaged and disengaged by an electric solenoid/hydraulic valve that operates a multiple disc clutch. A mechanical jaw clutch disengages the drive shaft from gears when not in use.

161. REMOVE AND REINSTALL. The front pto unit can be removed from the front of tractor as follows. Drain oil from unit by removing plug (1—Fig. 161) from the lower, center front of the housing.

Remove shaft cover if so equipped and disconnect oil lines (Fig. 161A) from pto housing. Disconnect the cable from lever (5—Fig. 161) and pull cable and housing free from pto unit. Secure lifting eyes and hoist to pto unit and remove the four attaching screws (4). Pull front pto assembly forward away from tractor.

When installing, carefully guide the drive shaft into the connecting sleeve at hydraulic pump. Tighten the front housing screws to 55 N·m (40 ft.-lbs.); the four retaining screws (4) to 400 N·m (300 ft.-lbs.) and the drain plug to 30 N·m (23 ft.-lbs.) torque. Fill unit

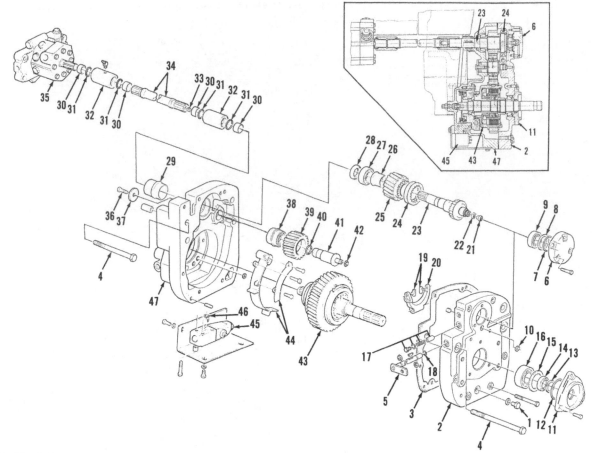

Fig. 161—Partially exploded view of front pto. The four larger screws (4) attach the unit to tractor.

1. Drain plug
2. Housing front half
3. Gasket
4. Attaching screws
5. Shift lever
6. Bearing quill
7. "O" ring
8. Shims
9. Tapered roller bearing
10. Seal ring
11. Bearing quill
12. "O" ring
13. Seal ring
14. Seal
15. Shims
16. Tapered roller bearing
17. Detent assembly
18. "O" ring
19. Guide shoes
20. Shift fork
21. Seal ring holder
22. Seal ring
23. Input shaft
24. Coupling
25. Input gear
26. Bushing
27. Tapered roller bearing
28. Seal
29. Plastic bushing
30. Spacer
31. "O" ring
32. Coupling
33. Spring
34. Drive shaft
35. Hydraulic pump
36. Cap screw
37. Washer
38. Bearing assembly
39. Intermediate gear
40. "O" ring
41. Intermediate shaft
42. "O" ring
43. Clutch & output shaft
44. Baffle
45. Control valve
46. Seal rings (4 used)
47. Housing rear half

with transmission oil and check for leaks, especially at tubing connections.

162. OVERHAUL. Most service to the front pto unit will necessitate removal of the unit as outlined in paragraph 161. Remove bearing quills (6 and 11—Fig. 161), being careful not to lose shims (8 and 15). Unbolt and remove the front half of the pto housing (2), remove plug and detent assembly (17), then lift out the input shaft (23) with drive gear (25), engagement collar (24) and bearings (9, 26 and 27). The intermediate gear (39), shaft (41) and bearings (38) can be removed from rear half of housing after removing retaining screw (36) and washer (37). Unbolt and remove oil baffle (44), then lift pto clutch and output shaft assembly (43) from rear housing and oil manifold. Oil manifold and brake assembly and control valve can be unbolted from housing.

Remove output shaft (81—Fig. 161C), hub (84) and associated parts. Remove backing plate (86), internally splined clutch friction discs (78) and plates (79) with external tabs. Use a special tool as shown in Fig. 161B to compress spring (75—Fig. 161C), then remove snap ring (77). Compressed air can be used to blow piston (73) from bore. Be careful to prevent ring (72) from catching in groove in drum (68) for snap ring (85). Remove snap ring (49), brake plate (50), disc (51) and plate (52), then use air to blow piston (53) from bore in manifold (63).

Inspect all parts and renew any that are grooved, cracked, discolored from heat or otherwise questionable.

Clutch plates (79), thickness new . . . . 2.3 mm
(0.09 inch)

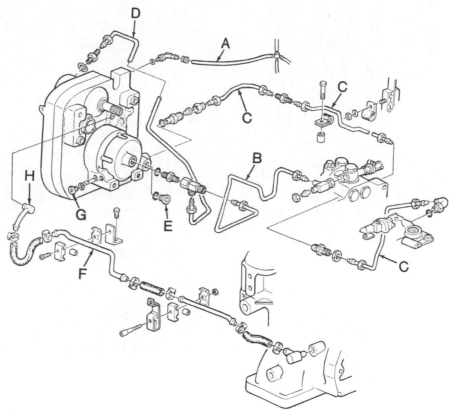

Fig. 161A—View of oil lines to front pto.

A. Leakage oil line
B. Lubrication oil line
C. Pressure oil line
D. Lubrication oil line
E. Pto clutch test port plug
F. Return line
G. Pto brake test port plug
H. Elbow fitting for return line

Clutch disc (78), thickness new . . . . . . 2.44-2.54 mm
(0.096-0.100 inch)
Brake disc (51), thickness new . . . . 5.9-6.1 mm
(0.232-0.240 inch)

Clutch spring (75)—
Free length . . . . .98 mm
(3.88 inches)
Tension at 46 mm . . . . .530-650 N
(1.80 inches)/(117-143 lbs.)

Modulation spring (60)—
Free length . . . . .87 mm
(3.42 inches)
Tension at 48 mm . . . . .53-65 N
(1.88 inches)/(12-14.5 lbs.)

Modulation spring (61)—
Free length . . . . .54 mm
(2.13 inches)
Tension at 31 mm . . . . .123-151 N
(1.22 inches)/(27.5-40 lbs.)

Assemble in reverse of disassembly procedure. Make sure the three seal rings (48) are in place in rear housing (47) before installing manifold and brake assembly, then tighten retaining screws to 55 N·m (40 ft.-lbs.) torque. Eight clutch plates (79) and eight discs (78) should be alternated, beginning with a plate (79) with external tangs and ending with a disc

(78) with internal splines. Backing plate (86) should be installed with flat side toward last friction disc (78). Screws attaching oil baffle (44—Fig. 161) should be coated with Loctite 242 to prevent loosening. Tighten screw (36) retaining shaft (41) to 75 N·m (55 ft.-lbs.) torque. Tighten screws holding halves of housing (2 and 47) together to 55 N·m (40 ft.-lbs.) and drain plug to 30 N·m (23 ft.-lbs.) torque. Bearings (9 and 27) and bearings (16—Fig. 161 and 64—Fig. 161B) should be adjusted to have 0.0-0.05 mm (0.0-0.002 inch) preload, by the addition of shims (8 and 15—Fig. 161). To determine the thickness of shims to

Fig. 161B—Use special tool JDT-24A or equivalent to compress spring so that snap ring can be removed or installed.

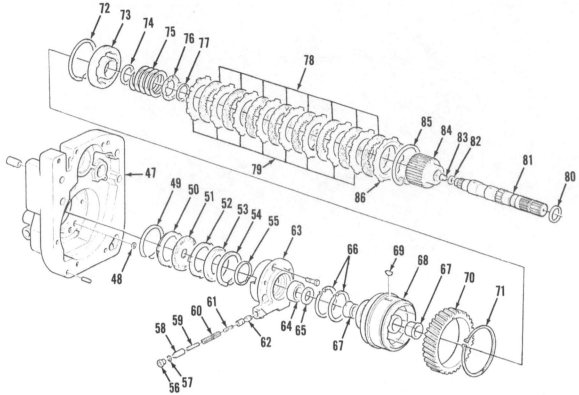

Fig. 161C—Exploded view of front pto clutch and brake.

47. Housing rear half	57. "O" ring	68. Clutch drum	78. Clutch friction discs
48. Seal rings (3 used)	58. Piston	69. Woodruff key	79. Clutch plates
49. Snap ring	59. Pin	70. Ring gear	(external lugs)
50. Brake plate	60. Modulator spring (long)	71. Snap ring	80. Back-up washer
51. Brake disc	61. Modulator spring (short)	72. Seal ring	81. Output shaft
(internal splines)	62. Modulation valve	73. Clutch piston	82. Seal ring
52. Brake plate	63. Oil manifold	74. Seal ring	83. Seal ring holder
53. Brake piston	64. Tapered roller bearing	75. Clutch release spring	84. Clutch hub
54. Seal ring	65. Spacer washer	76. Spring plate	85. Snap ring
55. Seal ring	66. Seal rings	77. Snap ring	86. Backing plate
56. Plug	67. Bushings		

install, push bearing races in to remove all play from bearings, then measure depth of bearing cup as shown at (B—Fig. 162). Measure the height of lip on appropriate quill as shown at (C). Subtract the height of the lip (C) from the depth of bearing (B), then add

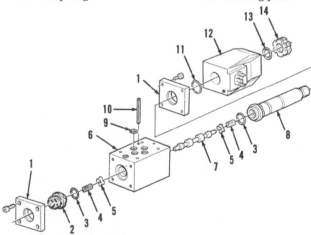

Fig. 163—Exploded view of solenoid valve used to operate front pto. Smaller cam of spool (7) should be toward end of housing (6) with spring pin (10).

1. Flange	8. Iron core
2. Cap	9. "O" rings (4 used)
3. "O" rings	10. Spring pin
4. Springs	11. Flange "O" ring
5. Centering discs	12. Solenoid
6. Housing	13. "O" ring
7. Spool	14. Slotted nut

Fig. 162—Refer to text and measure as shown to determine correct thickness of shims to be installed at (8 and 15-Fig. 161). Bearing cup is shown at (A).

the desired preload to this difference. Install shims (8 or 15—Fig. 161) equal to the calculated thickness. If removed, fitting (H—Fig. 161A) should be installed with a coating of Loctite 242 on the lip pressed into housing.

HYDRAULIC LIFT SYSTEM

The hydraulic lift system is a closed center, constant pressure type. The stand by pressure of 15800-16200 kPa (2300-2350 psi) is furnished by an eight piston, constant running variable displacement pump. Two sizes of main hydraulic pumps are available, a smaller pump displaces 22.6 cm³ (1.38 cubic inches) and a larger pump displaces 40 cm³ (2.4 cubic inches). Pump assemblies and internal parts will be different; however, operation of the two pumps is basically the same.

Pump is mounted in tractor front support and is driven by a coupling from front end of engine crankshaft. Tractors equipped with a front pto have a hydraulic pump with a through shaft that also drives the front pto unit. Charging oil for the hydraulic main pump is supplied by the transmission oil pump and oil not used by the main pump is routed to the auxiliary hydraulic oil reservoir. This reservoir provides an auxiliary supply of oil when transmission oil pump is unable to meet the demand of the main pump. When there is little or no demand by the main pump, the overflow from auxiliary reservoir is returned to the clutch housing through an oil cooler where part of it fills the brake control valve reservoir. The remainder lubricates the transmission shafts and gears.

TROUBLE-SHOOTING

All Models

165. The following are symptoms and their possible causes that may occur during the operation of the hydraulic lift system. By using this information in conjunction with the Test and Adjust information, no trouble should be encountered in servicing the hydraulic system.

1. Slow system operation. Could be caused by:
   a. Clogged transmission oil filter.
   b. Transmission oil pump inlet screen plugged.
   c. Faulty transmission oil pump.
   d. Transmission oil pump relief stuck open.
   e. Main hydraulic pump stroke control valve not seating properly.
   f. Oil leak on low-pressure side of system.
   g. Hydraulic pump shut-off screw turned in.
   h. Crankcase outlet valve seized.

2. Erratic pump operation. Could be caused by:
   a. Pump stroke control valve not seating properly.
   b. Leaking pump inlet or outlet valves or valve "O" rings.
   c. Broken or weak pump piston springs.

3. Noisy pump. Could be caused by:
   a. Worn drive parts or loose cap screws in drive coupling.
   b. Air trapped in oil cavity of pump stroke control valve.

4. No hydraulic pressure. Could be caused by:
   a. Pump shut-off valve closed.
   b. No oil in system.
   c. Faulty pump.

5. Rockshaft fails to raise or raises slowly. Could be caused by:
   a. Excessive load.
   b. Low pump pressure or flow.
   c. Rockshaft piston "O" ring faulty.
   d. Flow control valve maladjusted.
   e. Thermal relief valve or pressure relief valve defective.
   f. Cam follower adjusting screw maladjusted.
   g. Transmission oil filter plugged.
   h. Defective seals between cylinder and rockshaft housing or between rockshaft housing and transmission case.

6. Rockshaft settles under load. Could be caused by:
   a. Leaking discharge valve.
   b. Leaking rockshaft cylinder check valve.
   c. Leaking cylinder end plug.
   d. Faulty rockshaft cylinder valve housing.
   e. Thermal relief valve or pressure relief valve leaking.

7. Erratic action of rockshaft control valves. Could be caused by:
   a. Control valves maladjusted.
   b. Rockshaft piston "O" ring faulty.
   c. Discharge valve leaking.
   d. Thermal relief valve or pressure relief valve leaking.

8. Rockshaft lowers too fast or too slow. Could be caused by:
   a. Rate-of-drop screw maladjusted.

b. Valve linkage damaged.

9. Rockshaft raises too fast. Could be caused by:
   a. Flow control valve incorrectly set.

10. Insufficient load response. Could be caused by:
    a. Control valve clearance excessive.
    b. Control valves sticking.
    c. Control lever not positioned correctly on quadrant.
    d. Worn load control shaft or bushings.
    e. Negative stop screw turned in too far.

11. Hydraulic oil overheating. Could be caused by:
    a. Control valves adjusted too tight and held open.
    b. Control valves leaking.
    c. Control valve "O" rings faulty.
    d. Thermal relief valve or pressure relief valve leaking.
    e. Oil cooler plugged.

HYDRAULIC SYSTEM TESTS

Before making any tests on the main hydraulic pump or lift system, be sure the transmission oil pump is satisfactory as the performance of main pump is dependent upon being charged by the transmission oil pump. For information on testing of transmission oil pump, refer to paragraphs 124 and 131.

All Models

166. MAIN HYDRAULIC PUMP TEST. The main hydraulic pump should be tested for standby pressure as described in paragraph 168 before testing the rate of flow. Low standby pressure will result in low flow. Refer to paragraph 167 for flow testing procedure using hydraulic test unit after making sure that standby pressure is correct. Refer to paragraph 169 for a quick test procedure using a hydraulic test

unit attached to the selective control valve quick disconnect ports.

167. When testing the rate of flow of the main hydraulic pump using the hydraulic test unit, attach the unit as follows. Disconnect outlet line from pump in front of pressure control valve and connect the flow meter inlet line as shown in Fig. 170. Attach outlet from tester to the filter cover as shown or to the right side port of the left-hand quick coupler. If selective control valve port is used, the handle of the left-hand selective control valve must be secured in the rearward position during test.

Start engine and operate until transmission oil is heated to 65° C (150° F), then set engine speed at 2000 rpm. Close tester control valve so tester pressure gage reads 13800 kPa (2000 psi) and observe rate of flow which should be as follows:

2150, 2255 models —
   23 cm^3 (1.38 cu. in.) pump . . . . . . .34 L/m (9 gpm)

2155, 2355, 2355N, 2555 —
   23 cm^3 (1.38 cu. in.) pump . . . . . . .34 L/m (9 gpm)
   40 cm^3 (2.4 cu. in.) pump . . . . . . .68 L/m (18 gpm)

2350, 2550 models—
   23 cm^3 (1.38 cu. in.) pump . . . . . 21 L/m (5.5 gpm)
   40 cm^3 (2.4 cu. in.) pump . . . . . . .68 L/m (18 gpm)

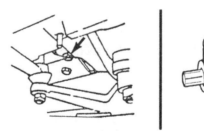

Fig. 172—Arrow indicates locations of pump stroke control valve adjuster screw and locknut.

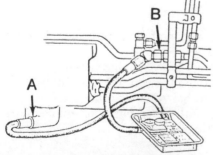

Fig. 170—A flow control and pressure test unit can be attached as shown in place of fitting which supplies system pressure to power steering system.

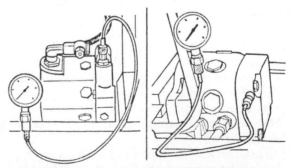

Fig. 173—View of gage attached for checking main pump standby pressure.

If main hydraulic pump will not meet both pressure and flow requirements, it must be removed and overhauled as outlined in paragraph 176.

168. Test main hydraulic pump standby pressure by installing a 35000 kPa (5000 psi) test gage as shown in Fig. 173.

Start engine and warm transmission oil to a temperature of 65° C (150° F). With engine running at 2000 rpm and transmission in NEUTRAL, move left selective control valve lever forward and note gage reading, which should be 15900-16200 kPa (2300-2350 psi). If pump standby pressure is incorrect, refer to Fig. 172, loosen jam nut and turn pump stroke control valve adjusting screw in to increase or out to decrease pressure. If pump will not produce the correct standby pressure, or if pressure pulsates more than 690 kPa (100 psi), the following are possible causes.

a. Gage improperly dampened.
b. Stroke control valve stuck or held open.
c. Crankcase valve open.
d. Pumping pistons scored.
e. Inlet or discharge valves leaking.
Refer to paragraph 176 for overhauling pump.

169. FLOW CONTROL VALVE TEST. A flow control valve is incorporated into the rockshaft valve

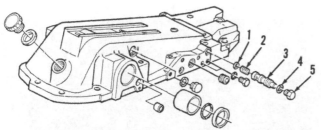

Fig. 175—View showing location of flow control valve in rockshaft housing.

1. Washers
2. Spring
3. Flow control valve
4. Seal ring
5. Plug

Fig. 176—View of negative stop screw and locknut.

circuit to reduce the oil flow of some models. See Fig. 175. Check the flow control valve setting with a hydraulic test unit as follows: Connect test unit inlet line to the selective control valve. Attach outlet line from tester to the oil filter relief valve bore. Open the test unit and run engine at 1000 rpm. Move selective control valve lever forward and close tester control valve until tester pressure gage shows 12000 kPa (1750 psi). Raise rockshaft and then check flow meter gage, which should indicate the following:

2150, 2255 models—
  23 cm^3 (1.38 cu. in.) pump . . . . . 21 L/m (5.5 gpm)

2155, 2355, 2355N, 2555—
  23 cm^3 (1.38 cu. in.) pump . . . . 19 L/m (5 gpm)
  40 cm^3 (2.4 cu. in.) pump . . . . 34 L/m (9 gpm)

2350, 2550 models—
  23 cm^3 (1.38 cu. in.) pump . . . . . 21 L/m (5.5 gpm)
  40 cm^3 (2.4 cu. in.) pump . . . . . . 36 L/m (9.5 gpm)

If flow is incorrect, refer to Fig. 175, remove plug, valve and spring, then vary washers (shims) as necessary. Also, check spring, which should test 55-65 N (11.7-14.3 lbs.) when compressed to a length of 20 mm (0.79 inch). This check also can be used as a quick check of the complete system.

HYDRAULIC SYSTEM ADJUSTMENTS

The following paragraphs outline the adjustments that can be made when necessary to correct faulty hydraulic operation, or that must be made when reassembling a hydraulic lift system that has been disassembled for service.

However, because of component parts' interaction, adjustments must be made in the following order:
  1. **Negative stop screw.**
  2. **Rockshaft control lever neutral range.**
  3. **Control lever position.**
  4. **Load control.**
  5. **Rate-of-drop.**

All Models

170. NEGATIVE STOP SCREW. The negative stop screw is shown in Fig. 176 and is used to provide a stop for the load control arm. To adjust, loosen jam nut and turn stop screw in until it just contacts the load control arm, then turn screw out the specified amount and tighten lock nut. Contact of stop screw with load control arm can be more easily felt if rockshaft housing filler plug is removed and a screwdriver is held against upper end of the load control arm. After screw just contacts arm, back screw out ⅛ turn on 2150, 2255, 2350, 2550 models, ⅓-½ turn on 2155, 2355, 2355N and 2555 models. It is important that this is correctly adjusted because it can adversely affect remaining system adjustments.

171. CONTROL LEVER NEUTRAL RANGE.

To adjust the control lever neutral range, load lift links with a weight of approximately 100 kg (220 lbs.). Remove the pipe plug, located directly in front of control lever tube, which will expose the control valve adjusting screw (1—Fig. 177 or Fig. 178). The adjusting screw is located under the Sound Gard Body of models so equipped. Move load selector lever (2) to "MIN" depth position and lower the lift arms fully. Move the rockshaft control lever (3) to center of quadrant or console and close the rate-of-drop screw. Start engine and run at 1300 rpm. When engine starts, turn adjusting screw (1) counterclockwise until the arms just start to raise, then turn adjusting screw (1) ¼ turn clockwise. Open the rate-of-drop screw. Move the control lever (3) forward until rockshaft just starts to lower and mark this point on edge of quadrant or console. Now move lever slowly rearward until rockshaft just begins to raise and mark this point on edge

of quadrant or console. Distance between these two marks should be as follows:

2150, 2255 . . . . 5 mm (0.2 inch)
2155 . . . . 2-4 mm (0.08-0.16 inch)
2355, 2555—
 Without Sound Gard
 Body . . . . 2-4 mm (0.08-0.16 inch)
 With Sound Gard Body . . . . 15 mm (0.6 inch)
2355N . . . . 3-6 mm (0.12-0.24 inch)

If not, turn the control valve adjusting screw (1) clockwise to increase or counterclockwise to decrease control lever neutral range.

172. CONTROL LEVER POSITION.

To adjust control lever position on models without Sound Gard Body, operate engine at 1300 rpm and place load selector (2—Fig. 178) in lower "MIN" position. Then, move rockshaft control lever (3) fully forward to completely lower rockshaft. Loosen control lever adjusting nut (4—Fig. 177) and move control lever rearward until there is the correct amount of clearance at (A) between control lever and end of quadrant slot. Rotate control shaft arm counterclockwise to the point where rockshaft just starts to raise. Tighten control lever nut (4). Correct clearance (A) is 10-25 mm (0.4-0.8 inch) for 2150, 2255, 2355N, 2350 and 2550 models; 10-13 mm (0.39-0.51 inch) for 2155, 2355 and

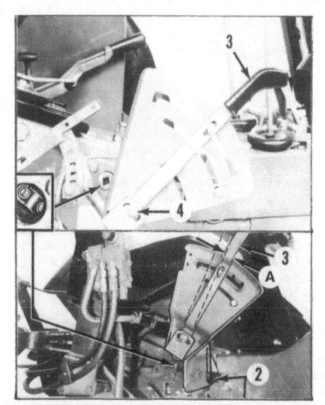

Fig. 177—Refer to text for adjustment of control lever neutral range.

1. Adjusting screw
2. Load selector lever
3. Rockshaft control lever
4. Adjusting nut

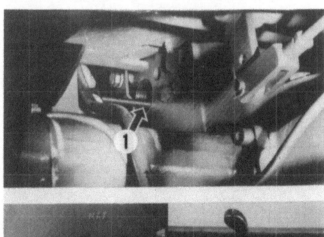

Fig. 178—View showing adjustment of control lever position on models without Sound Gard Body. Refer to text.

2555 models. Pull lever (3) to the rear and check to be sure that a small gap exists at the rear.

On models equipped with Sound Gard Body, remove side cover from console. Pull back upholstery, remove plastic rivets and remove plastic cover from right lift arm. Place load selector lever in "MIN" position. With engine operating at 1300 rpm, move control lever forward until rockshaft is fully lowered; then, move control lever to the rear until its front edge is aligned with position "7-7.5" on console. See Fig. 179. Working through the uncovered opening, adjust position of rod end (B—Fig. 179A) on link (A) until rockshaft starts to raise. Tighten locknuts on turnbuckle. Move control lever rearward until its front edge is aligned with "1.5-2.5" position. Rockshaft should now raise to its highest position. If incorrect, readjust at "7-7.5" position.

173. LOAD CONTROL ARM. On models without Sound Gard Body, remove rockshaft filler hole plug to expose the load control arm cam follower adjusting screw (5—Fig. 180). Place load selector lever (2—Fig. 177) in "MAX" position. With engine operating at 1300 rpm, move control lever (3) fully forward to completely lower rockshaft. Then, move control lever rearward until distance (A) between control lever and end of quadrant slot is correct. Distance (A) should be 85-95 mm (3.34-3.74 inches) for 2150, 2255, 2350 and 2550 models; 47-53 mm (1.88-2.12 inches) for 2155, 2355 and 2555 models; 85-100 mm (3.34-3.94 inches) for 2355N model. Hold control lever in this position, loosen jam nut and turn cam follower adjusting screw

(5—Fig. 180) counterclockwise until rockshaft is fully lowered. Then, turn adjusting screw clockwise until rockshaft starts to raise. Tighten jam nut. A special tool (JDG402) is available to facilitate adjustment.

> **NOTE: If rockshaft begins to raise before control lever reaches the recommended setting, it will be necessary to turn cam follower adjusting screw counterclockwise to allow the lever to be positioned while rockshaft remains in lowered position.**

On models equipped with Sound Gard Body, remove rockshaft filler plug. Place load selector lever in "MAX" position. With engine operating at 1300 rpm, move control lever forward until rockshaft is fully lowered. Move control lever rearward until its front edge is aligned with position "2-2.5" (Fig. 179). If rockshaft starts to raise before control lever reaches the "2.5" position, loosen jam nut and turn cam follower adjusting screw (5—Fig. 180) counterclockwise to allow lever to be positioned while rockshaft remains in lowered position. Turn adjusting screw (5) clockwise until rockshaft just starts to raise, then tighten jam nut.

174. RATE-OF-DROP. The rate-of-drop adjusting screw is located on top side of rockshaft housing. Fig.

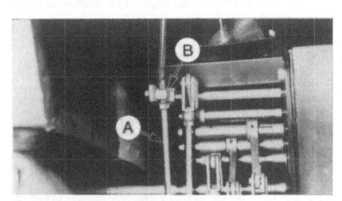

Fig. 179A—Refer to text when adjusting length of link (A) by repositioning end (B).

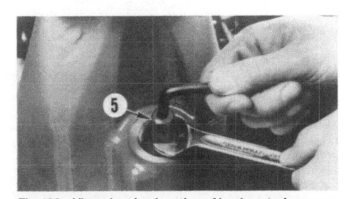

Fig. 180—View showing location of load control arm cam follower adjusting screw (5). Remove rockshaft filler plug for access.

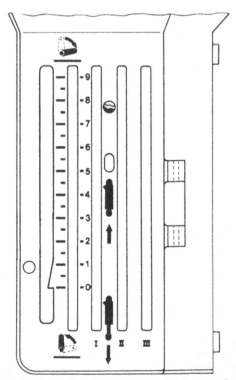

Fig. 179—Top view of hydraulic control console on models with Sound Gard Body.

Illustrations for Fig. 179, Fig. 179A and Fig. 180 reproduced by permission of Deere & Company. Copyright Deere & Company.

181 shows adjusting screw with handle (6) for models with and without Sound Gard Body.

On all models, turn adjusting screw clockwise to decrease rate-of-drop or counterclockwise to increase rate-of-drop. Tighten jam nut after adjustment is completed. Rate-of-drop will vary with the weight of the attached implement. Rate-of-drop time should not be less than two seconds. However, adjusting screw should not be turned out more than one complete turn.

MAIN HYDRAULIC PUMP

All Models

175. A 23 cc (1.38 cu. in.) or a 40 cc (2.40 cubic inch) piston type pump is used. Pumps with a drive shaft that continues through pump are available to drive the front pto unit.

Pump should maintain about 14000 kPa (2050 psi) working pressure with a standby pressure of 15900-16200 kPa (2300-2350 psi). Standby operation of the pump occurs when pressure in the pump crankcase builds high enough to hold the pump pistons away from the pump cam. Pump crankcase (standby) pressure is controlled by the stroke control valve located in a bore in the pump valve housing.

A pump shut-off (destroking) screw may be optionally available which will make pump inoperative and will act as an aid during cold weather starts.

176. R&R AND OVERHAUL. Remove grille screens, disconnect light bar wires at connector and remove hood. Drain fuel from main fuel tank, lines and remote tank, then disconnect lines from fuel tank. Remove air cleaner and air conditioning condenser if so equipped. Remove fuel tank brackets and lift fuel tank from tractor. On models with through drive, it is necessary to remove the radiator. On all models, disconnect oil lines at pump and immediately plug or cap all openings to prevent loss of oil and dirt from entering system. Remove the four screws (4—Fig. 182 or 4A—Fig. 183) and loosen the clamp screw (2—Fig. 182) or set screws (2A—Fig. 183). Slide pump drive adapter (3, 3H, 3P or 3PH—Fig. 182 or Fig. 183)

Fig. 181—View of rate-of-drop adjustment screw (6) for models with and without Sound Gard Body. Refer to text.

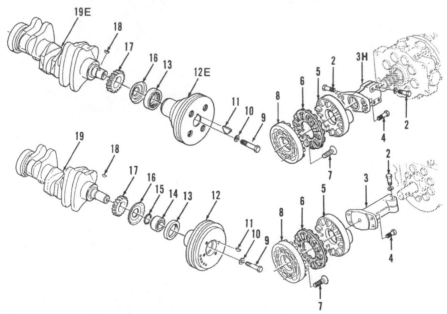

Fig. 182—Exploded view of pump drive and crankshaft used on models without front pto. Note differences for models with standard and high volume hydraulic pumps.

1. Hydraulic pump	5. Coupling	11. Shaft key
2. Clamp screw	6. Cushion	12. Pulley for later models
3. Pump drive shaft	7. Allen screw	12E. Pulley for early models
3H. Drive shaft for high volume pump	8. Adapter coupling	13. Oil seal
4. Cap screws	9. Center screw	14. Seal sleeve
	10. Washer	15. "O" ring
		16. Oil slinger
		17. Crankshaft timing gear
		18. Drive key
		19. Late crankshaft
		19E. Early crankshaft

forward on pump shaft. Remove drive cushion (6, 6P or 6PH) and front coupler half (5—Fig. 182). Unbolt and remove pump assembly.

176A. OVERHAUL 23 cc (1.38 cu. in.) PUMP. Thoroughly clean exterior of pump and check pump shaft end play before disassembling. Use a dial indicator to accurately measure pump shaft end play and record measurement for aid in reassembling. End play should be 0.1-0.9 mm (0.004-0.035 inch) for models without through shaft, 0.025-0.10 mm (0.001-0.004 inch) for models with through drive shaft for the front pto. If end play is excessive, install new thrust washers (21—Fig. 184) when assembling models without through shaft. To reduce end play on models with through shaft, add shims (57) as necessary when assembling to reduce end play to within limits. Bearing wear or wrong number of adjusting shims can cause excessive end play.

Clamp pump in a vise and remove cover (2), then remove crankcase outlet valve (34, 40, 43 and 49) from pump housing. Remove outer thrust washer (21).

NOTE: At this time, shaft (23) can be removed if desired by removing Woodruff key (50) and pushing shaft out of body and cam ring (22). Be sure not to lose any of the 33 rollers (24) that will be loose. However if pump requires disassembly, it is a good policy to completely disassemble pump and inspect all parts and to remove pump pistons before removing pump shaft.

Remove piston assemblies (20, 46, 47 and 48), then if not already removed, remove shaft (23), rollers (24)

and cam ring (22). Remove inlet valves (8, 10, 11 and 12) and outlet valves (14, 15, 53, 17, 18 and 20). Identify all pistons, valves, springs and seats so they can be reinstalled in their original positions. Remove plug (42) and filter screen (45). Loosen jam nut and remove stroke control valve adjusting screw (29), spring (31), spring guide and valve (32).

Clean and inspect all parts. Use the following specifications as a guide for renewal of parts.

Thrust washer (21) thickness . . . . 2.21-2.31 mm
                                (0.087-0.091 inch)
Piston bore ID . . . . 17.27-17.30 mm
                                (0.680-0.681 inch)
Piston (46) OD . . . . 17.25-17.27 mm
                                (0.679-0.680 inch)
Cam race ID . . . . 45.72-45.75 mm
                                (1.800-1.801 inch)
Pump shaft (23) cam OD . . . . 37.77-37.79 mm
                                (1.487-1.488 inch)

Discard all old "O" rings and use new during assembly. Pay particular attention to shaft bore around the quad ring seal groove as leakage at this point can cause the pump to be slow in going out of stroke. If outlet valve seats (14) are damaged, drive them out and install new seats with large chamfered end toward bottom of bore. Press seats in to 29.75 mm (1.171 inch) below spot face surface of bore using special tool JDH-39-1 or equivalent. OD of new outlet valve (15) is 15.47-15.5 mm (0.609-0.611 inch) and any valve that is distorted, scored or worn should be renewed. If stroke control valve seat (33) is damaged, remove plug (20) or shut-off assembly (35) if so

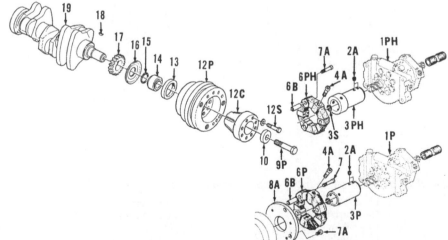

Fig. 183—Exploded view of pump drive and crankshaft used on models with front pto. Note differences for models with standard and high volume hydraulic pumps.

1P.	Hydraulic pump with through shaft	4A. Allen screws	8A. Adapter
2A.	Set screws	6B. Bushings	9P. Center screw
3P.	Pump drive shaft	6P. Coupling	10. Washer
3PH.	Drive shaft for high volume pump	6PH. Coupling for high volume pump	12C. Coupling
3S.	Snap ring	7. Allen screws	12P. Pulley
		7A. Allen screws	12S. Coupling screws
			13. Oil seal

14. Seal sleeve
15. "O" ring
16. Oil slinger
17. Crankshaft timing gear
18. Drive key
19. Crankshaft

equipped and drive out seat (33). Install new valve chamfered end first and drive it in bore until it bottoms. Spring (31) should test 710-850 N (160-190 lbs.) when compressed to a length of 63.5 mm (2½

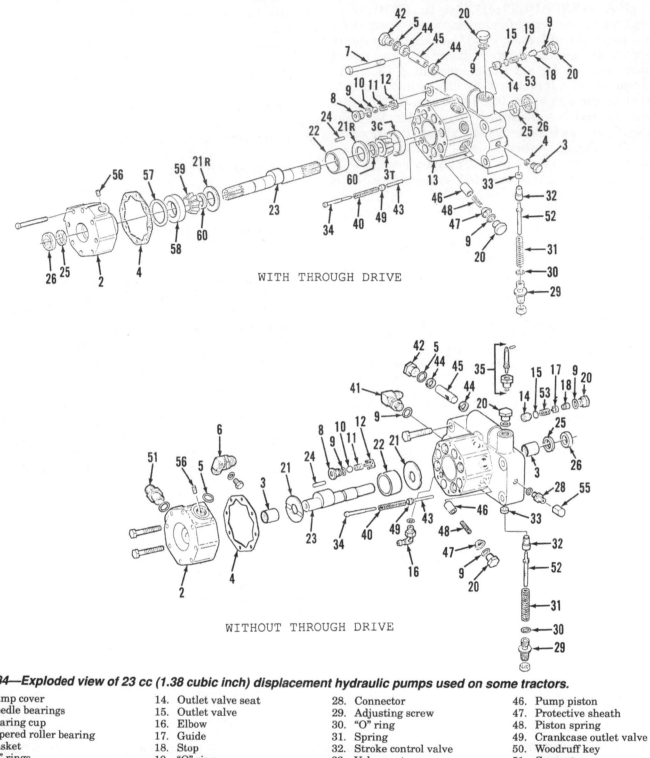

Fig. 184—Exploded view of 23 cc (1.38 cubic inch) displacement hydraulic pumps used on some tractors.

2. Pump cover	14. Outlet valve seat
3. Needle bearings	15. Outlet valve
3C. Bearing cup	16. Elbow
3T. Tapered roller bearing	17. Guide
4. Gasket	18. Stop
5. "O" rings	19. "O" ring
6. Elbow	20. Plug
7. Screws	21. Thrust washer
8. Inlet valve seat	21R. Thrust washer
9. "O" ring	22. Cam race
10. Inlet valve ball	23. Pump shaft (cam)
11. Spring	24. Bearing rollers (33)
12. Guide	25. Quad ring
13. Housing	26. Oil seal

28. Connector	46. Pump piston
29. Adjusting screw	47. Protective sheath
30. "O" ring	48. Piston spring
31. Spring	49. Crankcase outlet valve
32. Stroke control valve	50. Woodruff key
33. Valve seat	51. Connector
34. Spring guide	52. Valve guide
35. Pump shut-off assy.	53. Spring
40. Spring	55. Cap
41. Elbow	56. Orifice
42. Plug	57. Spacer
43. Pin	58. Bearing cup
44. Seals	59. Tapered roller bearing
45. Filter screen	60. Spacer

inches). Renew any pistons (46) which are scored or pitted. Piston springs (48) should exert 80-100 N (18-22 lbs.) when compressed to 32 mm (1.26 inches) and springs must test within 7 N (1.5 lbs.) of each other when compressed to test length of 32 mm (1.26 inches).

When reassembling, dip all parts in oil. Use grease to hold roller in ID of cam ring during assembly. Seal (26) is installed with printed side toward outside. Thrust washers are installed with grooved side away from pump shaft cam.

Cover to body screws—
   First torque . . . . .20 N•m (14 ft.-lbs.)
   Final torque . . . . .50 N•m (35 ft.-lbs.)
Torque, piston plugs (20) . . . . . .125 N•m (90 ft.-lbs.)
Torque, pump mounting screws . . . . 120 N•m
                                          (85 ft.-lbs.)

Reinstall pump by reversing the removal procedure and adjust standby pressure as outlined in paragraphs 167 and 168.

176B. OVERHAUL 40 cc (2.4 cu. in.) PUMP. Thoroughly clean exterior of pump and check pump shaft end play before disassembling. Use a dial indicator to accurately measure pump shaft end play and record measurement for aid in reassembling. End play should be 0.025-0.10 mm (0.001-0.004 inch) for all models. End play is adjusted by adding or removing shims (26—Fig. 187). Bearing wear or wrong number of adjusting shims can cause excessive end play.

To disassemble the pump, remove the four cap screws retaining stroke control housing (15—Fig. 188) to front of pump and remove housing. Mark pump housing and place removed parts in a compartmented tray so parts will be reinstalled in their original locations. Withdraw outlet stops, springs, guides and valves (28 through 31—Fig. 187). Be especially careful to mark the pistons and bores so that each can be reinstalled in same bore from which it was removed. Remove all piston plugs (35), springs (34) and pistons (33), then carefully withdraw pump shaft (22) with bearing cones (19 and 24), thrust washers (21), spacers (20), cam race (23) and bearing needles (27).

Remove plug (7) retaining inlet valve assembly (5) and check inlet valve lift using a dial indicator. Lift should be 2.0-3.0 mm (0.080-0.120 inch). If lift exceeds 3 mm (0.120 inch), spring retainer is probably worn and valve should be renewed. Also check for apparent excessive looseness of valve stem in guide. Do not remove inlet valve assembly unless renewal is indicated or outlet valve seat (32) must be renewed. Do not reinstall a removed inlet valve (5) because they depend on their tight press fit for sealing. To remove an inlet valve, use a small pin punch and drive valve out, working through outlet valve seat (32). Outlet

valve seat can be driven out after inlet valve is removed. Outlet valve spring (30) should test 11.3-14.0 N (2.5-3.1 lbs.) when compressed to a length of 7.6 mm (0.30 inch).

All eight piston springs (34) should be of the same color and should test within 7 N (1.5 lbs.) of each other when compressed to length of 41 mm (1.62 inches). Piston springs should test as follows when compressed to 41 mm (1.62 inches).

Yellow . . . . .151-158 N (34.0-35.5 lbs.)
Green . . . . .158-165 N (35.5-37.0 lbs.)
Blue . . . . .165-171 N (37.0-38.5 lbs.)
Red . . . . .171-178 N (38.5-40.0 lbs.)

Inspect bearings (19 and 24), spacers (20), thrust washers (21), shaft (22), race (23) and bearing needles (27) for excessive wear or other damage and renew as necessary.

When reassembling pump, install new seal ring (10) in housing on early style pump and on shaft on late pump. Install oil seal (9) only deep enough to allow snap ring to enter groove, to avoid blocking the relief hole in body.

Valves located in housing (15—Fig. 188 or Fig. 188A) control pump output as follows: The closed

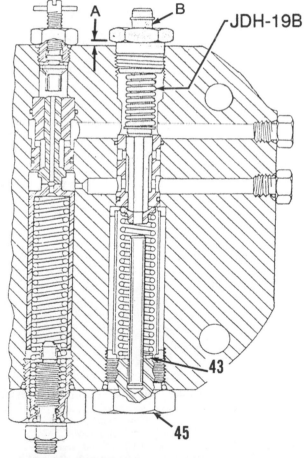

Fig. 186—Special adjusting tool (JDH-19B) can be used to determine stroke control valve setting. Refer to text.

hydraulic system has no discharge except through the operating valves or components. Peak pressure is thus maintained for instant use. Pumping action is halted when line pressure reaches a given point by pressurizing the camshaft reservoir of pump housing, thereby holding pistons outward in their bores.

The cut-off point of pump is controlled by pressure of spring (27) and can be adjusted by turning adjusting screw (32). Adjustment procedure is given in paragraphs 166, 167 and 168. When pressure reaches the standby setting, valve (23) opens and meters the required amount of fluid at reduced pressure in crankcase section of pump. Crankcase outlet valve (18—Fig. 188) or (34—Fig. 188A) is held closed by hydraulic pressure and blocks the outlet passage. When pressure drops as a result of system demands, crankcase outlet valve is opened by pressure of spring (19) and a temporary hydraulic balance on both ends of valve, dumping the pressurized crankcase fluid and pumping action resumes.

Cut-in pressure on late style pump is determined by pressure of spring (19—Fig. 188). Spring (19) should test 63-77 N (14-17 lbs.) when compressed to a length of 74.5 mm (3 inches).

Cut-in pressure on early style pump is determined by thickness of shim pack (43—Fig. 188A) and pressure of spring (19). A special tool JDH-19B is available to determine thickness of shim pack. See Fig. 186. Assemble outlet valve (34—Fig. 188A) and all components (19, 35, 36, 37, 38, 39, 40, 42, 44 and 45) using existing shim pack (43). Install special tool in place of plug (46) threading special tool into housing until gap (A—Fig. 186) is 3 mm (1/8 inch) as shown. If shim pack thickness is correct, scribe line on tool plunger should align with edge of tool plug bore (B). If not, remove stop plug (45) and add or remove shim washers (43) as required. Shims (43) are available in 0.25 and 0.76 mm (0.010 and 0.030 inch) thicknesses. If special tool JDH-19B is not available, use shim pack of same thickness as those removed, then add shims to raise cut-in pressure or remove shims to lower pressure.

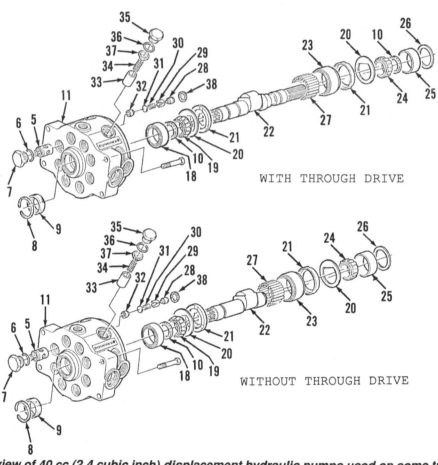

WITH THROUGH DRIVE

WITHOUT THROUGH DRIVE

Fig. 187—Exploded view of 40 cc (2.4 cubic inch) displacement hydraulic pumps used on some tractors.

5. Inlet valve (8)	18. Bearing cup	25. Bearing cup	32. Seat
6. "O" ring	19. Bearing cone	26. Shim	33. Piston (8)
7. Plug	20. Spacers	27. Bearing needles (25)	34. Spring
8. Snap ring	21. Thrust washers	28. Stop	35. Plug
9. Oil seal	22. Pump shaft (cam)	29. Guide	36. "O" ring
10. Seal ring	23. Race	30. Spring	37. Sheath
11. Housing	24. Bearing cone	31. Outlet valve (8)	38. Packing ring

On both early and late style pumps, add or remove shims (26—Fig. 187) as necessary to obtain pump shaft end play of 0.025-0.100 mm (0.001-0.004 inch). Renew all "O" rings, packings and seals and lubricate all parts with clean hydraulic oil. Install the stroke control housing (15-Fig. 188 or Fig. 188A) on pump housing making sure that oil passages for both seal rings (7) are aligned. **Do not attempt to use an early-type housing (15—Fig. 188A) with a late-type pump or a late-type housing (15—Fig. 188)** with an early-type pump. Bolt holes will align, but high-pressure passages will NOT and damage or injury may result. Tighten stroke control valve housing (15— Fig. 188A) retaining cap screws, plugs and other screws to the proper torque as follows.

Early pump (Without Serial No.)

Housing (15—Fig. 188A) to body screws—
    First torque . . . . 40-68 N•m (30-50 ft.-lbs.)
    Final torque . . . . .115 N•m (85 ft.-lbs.)
Test port plug
    (17—Fig. 188A) . . . . .34 N•m (25 ft.-lbs.)
Adjusting screw locknut
    (33—Fig. 188A) . . . . .60 N•m (45 ft.-lbs.)
Inlet valve plug
    (7—Fig. 187) . . . . .136 N•m (100 ft.-lbs.)

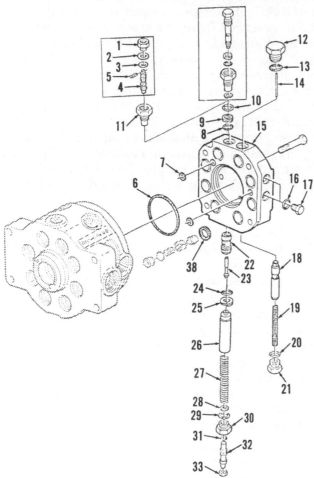

Fig. 188—Exploded view of stroke control valve assembly used on some 40 cc (2.4 cubic inch) displacement pump.

1. Bushing	
2. "O" ring	
3. "O" ring	18. Outlet valve
4. Lock out screw	19. Spring
5. Spring pin	20. "O" ring
6. "O" ring	21. Plug
7. Packing ring	22. Sleeve
8. "O" ring	23. Control valve
9. Washer	24. "O" ring
10. "O" ring	25. Back-up ring
11. Bushing	26. Guide
12. Plug	27. Spring
13. "O" ring	28. Washer
14. Pin	29. "O" ring
15. Housing	30. Bushing
16. "O" ring (2)	31. "O" ring
17. Plug (2)	32. Adjusting screw
	33. Jam nut

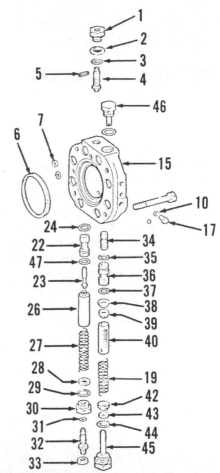

Fig. 188A—Exploded view of pump cover and stroke control valve assembly used on some 40 cc (2.4 cubic inch) displacement pumps. Refer to text for differences between this type and that shown in Fig. 188. Refer to Fig. 188 for legend except the following.

34. Crankcase outlet valve	40. Filter	
35. "O" ring	42. Seal (same as 38)	
36. Crankcase outlet valve sleeve	43. Shims	
37. "O" ring (same as 47)	44. "O" ring	
38. Seal (same as 42)	45. Plug	
39. Spring guide	46. Plug	

Piston plugs
(35—Fig. 187) . . . . .140 N•m (100 ft.-lbs.)
Pump drive to pump
shaft screws . . . . .50 N•m (35 ft.-lbs.)
Pump mounting screws . . . . . . .120 N•m (85 ft.-lbs.)

Late pump (With Serial No.)
Housing (15—Fig. 188)
to body screws . . . . .120 N•m (88 ft.-lbs.)
Test port plug
(17—Fig. 188) . . . . 34 Nm (25 ft.-lbs.)
Adjusting screw locknut
(33—Fig. 188) . . . . .60 N•m (45 ft.-lbs.)
Piston plugs
(35—Fig. 187) . . . . .185 N•m (135 ft.-lbs.)
Pump drive to pump
shaft screws . . . . .50 N•m (35 ft.-lbs.)
Pump mounting screws . . . . . . .120 N•m (85 ft.-lbs.)

On all models, reinstall pump by reversing the removal procedure and adjust standby pressure as outlined in paragraphs 167 and 168.

ROCKSHAFT HOUSING, CYLINDER AND VALVE

All Models

177. REMOVE AND REINSTALL. On models without Sound Gard Body, disconnect battery ground straps, disconnect wires from starter safety switch and remove seat assembly. Disconnect lift links from rockshaft arms. Disconnect return line to rockshaft housing and lines from selective control valves to remote couplers, then remove the breakaway couplers. Attach a hoist to rockshaft housing, place load selector lever in "MAX" position, then unbolt and lift off rockshaft housing assembly.

On models equipped with Sound Gard Body, remove Sound Gard Body as outlined in paragraph 186. Disconnect wires from starter safety switch. Disconnect lift links from rockshaft arms. Disconnect oil return line from selective control valves to rockshaft

and lines to remote couplers. Move shift rod of load selector lever upward (MAX position). Attach a hoist to rockshaft housing, then unbolt and lift off rockshaft assembly.

When reinstalling, use a new gasket and reverse the removal procedure. Tighten mounting cap screws to a torque of 120 N•m (85 ft.-lbs.) in the sequence shown in Fig. 189.

178. OVERHAUL. With rockshaft housing assembly removed, disassemble as follows: If so equipped, unbolt and remove filler tube. Remove rockshaft arms from rockshaft. Remove selector lever linkage from load control arm. Remove adapter (6—Fig. 190) and the rate-of-drop adjusting parts (3, 4, 5, 6 and 7), then use a magnet to remove rate-of-drop ball (2) from hole. Remove front outside cylinder mounting cap screw (1), then turn unit so bottom side is accessible. Drive out spring pin (D—Fig. 189A) which retains control shaft to pivot block. Remove snap ring (A) and unhook spring (B). Unbolt plate from rockshaft housing and remove quadrant and control shaft. Remove remote cylinder outlet adapter from left side of housing and unhook valve spring from linkage. Remove the four remaining cylinder attaching cap screws (33—Fig. 190). Lift cylinder assembly from housing and disengage selector arm from roller link as cylinder is removed. If not previously removed, rate-of-drop ball (2) will fall.

Unbolt and remove cam (11—Fig. 192) from rockshaft, then pull rockshaft (4) from crank arm (5) and housing. Bushing (6—Fig. 191), "O" ring (5) and retainer (4) will be removed from the right side by the rockshaft as rockshaft is removed from right side. If necessary, separate piston rod (7—Fig. 192) from crank arm (5) by driving out spring pin (6). Load selector shaft can also be removed, if necessary. Remove flow control valve assembly from right front of hydraulic housing.

With components removed, disassemble cylinder and valve unit as follows: Remove plug (8—Fig. 190),

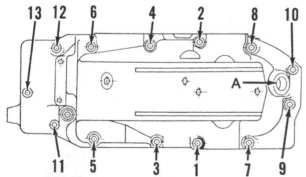

Fig. 189—Tighten screws attaching rockshaft housing to tractor in the sequence shown.

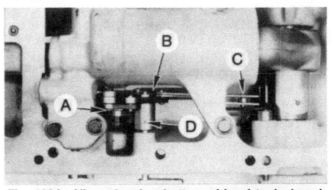

Fig. 189A—View showing bottom side of typical rockshaft housing. Spring pin (D) retains control lever shaft to pivot block. Spring (B) is also shown at (9—Fig. 193).

spring (10) and check ball (2). Remove surge pressure relief valve (12). Remove retainer ring and pull linkage from linkage pivot pin. Remove snap rings (32), plugs (30), springs (27), sleeves (26), valve spools (25) and seats (24) from bores in housing. Keep valves identified so they can be reinstalled in original bores. Remove piston from cylinder by bumping open end of cylinder against a wood block or remove plug (14) and push piston out. If white seal ring (10—Fig. 192) is renewed with the black pressure ring (9), wait about 10 minutes until seal ring has contracted before attempting to install piston.

Clean and inspect all parts for undue wear or other damage. Check valve spring (10—Fig. 190) should test 18-21 N (4.0-4.7 lbs.) when compressed to a length of 19 mm (0.75 inch). Control valve springs (27) should test 43-51 N (10.0-11.5 lbs.) when compressed to a length of 23 mm (0.91 inch). Flow control valve spring (24—Fig. 191) should test 55-65 N (11.7-14.3 lbs.) when compressed to a length of 20 mm (0.79 inch). Inspect the seating area of the two control

valves (25—Fig. 190) carefully. Leakage in this area could cause rockshaft settling if it occurs in the discharge (upper) valve, or upward rockshaft creep if it occurs in the pressure (lower) valve. Inspect seats (24) and sleeves (26) for any damage. Check rockshaft assembly for fractures, damaged splines or other damage. Check control linkage for worn or bent conditions. A worn adjusting cam (10—Fig. 193) will cause difficulty in adjusting lever neutral range.

When reassembling the rockshaft and cylinder unit, use sealant on threads of any pipe plugs that were removed and coat all "O" rings with oil. Start rockshaft into housing, align master splines of rockshaft (4—Fig. 192) and crank arm (5), then push rockshaft into position. Install rockshaft bushings (6—Fig. 191), which will be free in rockshaft bores, then install "O" rings (5) and retainers (4) with cupped side outward. Preassemble cam (11—Fig. 192A), pin (14), snap rings (1) and tube (15) with offset hole in tube (15) for pin (14) positioned as shown. Install cam (11—Fig. 192) on rockshaft (4) and

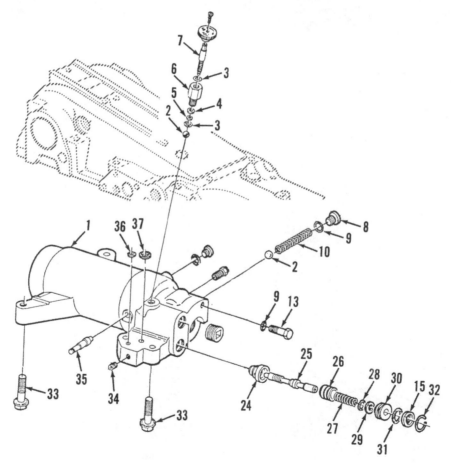

Fig. 190—Exploded view of rockshaft valve components.

1. Cylinder housing	8. Plug	24. Valve seat (2)	31. "O" ring (2)
2. Ball	9. "O" ring	25. Valve spool (2)	32. Snap ring (2)
3. "O" ring	10. Spring	26. Sleeve (2)	33. Cap screws (4)
4. Washer	12. Surge relief valve	27. Spring (2)	34. Plug
5. Snap ring	13. Cap screw	28. "O" ring (2)	35. Pivot pin
6. Adaptor	14. Plug	29. Back-up ring (2)	36. Seal ring
7. Rate-of-drop screw	15. Back-up ring	30. Plug (2)	37. Seal ring

tighten screw (12) to 15 N·m (10 ft.-lbs.) torque. Attach connecting rod (7) to control arm (5). Insert valve seats (24—Fig. 190) in their original bores, small end first, then place valve spools (25) in seats with smaller diameter of spools in seats. Install sleeves (26), chamfered end first, and valve springs (27). Install "O" rings (28) and back-up rings (29) in

bore of plugs (30) with back-up rings toward outer end. Install "O" rings (31) on outside of plugs, then carefully install plugs over ends of valves. Install

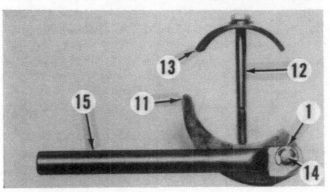

Fig. 192A—View of tube (15) with offset hole for pin (14) correctly attached to cam (11).

Fig. 191—Exploded view of typical rockshaft housing.

2.	Filler cap	27.	"O" ring
3.	Gasket	28.	Load selector lever
4.	Retainer	29.	Lever
5.	"O" ring	30.	Washer
6.	Bushing	31.	Nut
17.	Starter safety switch	32.	"O" ring
23.	Shim	33.	Bushing
24.	Spring	36.	Load control arm

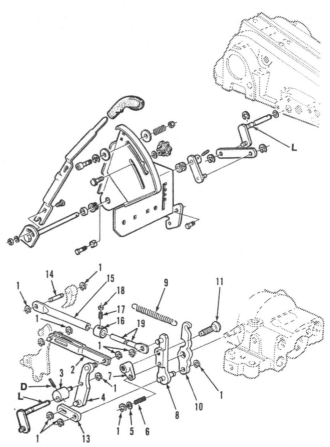

Fig. 193—Exploded view of control valve linkage. Spring (9) is also shown at "B" in Fig. 189A.

D.	Spring pin		9.	Spring
L.	Lever		10.	Adjusting cam
1.	Retaining rings		11.	Adjusting screw
2.	Roller link		13.	Link
3.	Pivot block		14.	Pin
4.	Link		15.	Tube
5.	Washer		16.	Collar
6.	Spring		17.	Set screw
7.	Special nut		18.	Locknut
8.	Link		19.	Rod

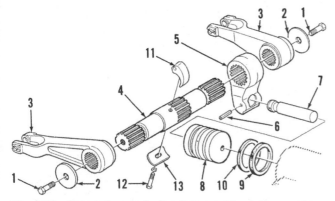

Fig. 192—View of rockshaft, piston and relative parts.

1.	Cap screw		
2.	Washer	8.	Piston
3.	Lift arm	9.	Pressure ring
4.	Rockshaft	10.	Sealing ring
5.	Crank arm	11.	Cam
6.	Spring pin	12.	Cap screw
7.	Piston rod	13.	Spacer

back-up rings (15) and snap rings (32). If white seal ring (10—Fig. 192) is renewed with the black pressure ring (9), wait about 10 minutes until seal ring has contracted before attempting to install piston. Lubricate piston assembly and install it in cylinder. Refer to Fig. 190 and install check valve ball (2), spring (10) and plug (8) with new "O" ring (9). Install surge pressure relieve valve (12). Assemble linkage and install on linkage pivot pin. Reinstall flow control valve and if not already done, remove rate-of-drop adjusting screw (7) from rockshaft housing. Install load selector control lever shaft (36—Fig. 191) in rockshaft housing. Install cylinder and valve assembly in rockshaft housing, while starting slot of roller link (2—Fig. 193) over selector control lever shaft, entering the connecting rod in piston and entering the rod (19—Fig. 193A) into tube (15). Use new seal rings and install and front cap screw (13—Fig. 190) and the

four screws (33). Install all five screws before first tightening front cap screw (13) to 70 N·m (50 ft.-lbs.), then tighten four remaining locking cap screws (33) to 70 N·m (50 ft.-lbs.) torque. Move the crank arm toward rear and hold a strip of metal 2-3 mm (0.08-0.12 inch) thick between crank arm (5—Fig. 193A) and inside of rockshaft housing as shown. Loosen lock nut (18) and set screw (17), slide collar (16—Fig. 193) against tube (15), then tighten set screw and lock nut to retain position of the collar on rod. Install linkage (2, 3 and 4—Fig. 193), making sure that pin on link (4) engages link (13) and install snap ring (A—Fig. 189A). Offset hole in pivot shaft (3—Fig. 193) should be down (with rockshaft inverted) when installing, then install pin (D). Connect the linkage spring and make sure that lever (L—Fig. 193) is correctly positioned. Turn rockshaft over to correct position, then install rate-of-drop ball (2—Fig. 190), adapter (6) and adjusting screw (7). Install control quadrant assembly and selector lever.

Reinstall rockshaft assembly by reversing the removal procedure and adjust as outlined in paragraphs 170 through 174.

LOAD CONTROL (SENSING) SYSTEM

All Models

179. The load sensing mechanism is located in the rear of transmission case. Refer to Fig. 194 and Fig. 194A for views showing component parts.

Load control shaft (9) is mounted in tapered bushings. As load is applied to the shaft ends from the hitch links, the middle of shaft flexes forward and

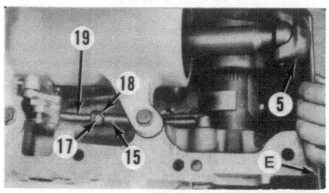

Fig. 193A—View into rockshaft housing showing method of adjusting position of collar (16—Fig. 193). Spacer (E) is shown held in position between crank arm (5) and housing.

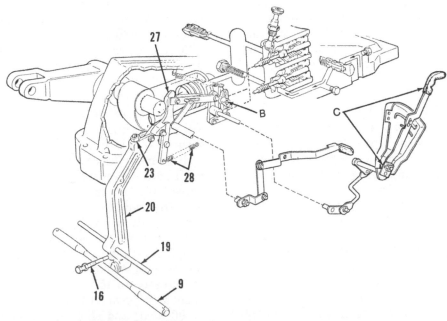

Fig. 194—Drawing of typical load control mechanism and rockshaft control parts. Refer to Fig. 194A for legend.

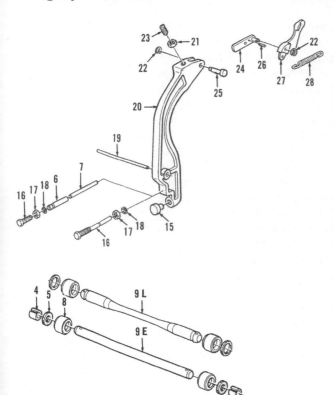

Fig. 194A—Load control mechanism is located in rear of transmission case. Flexing of control shaft (9) actuates load control arm (20).

4. Bushing
5. Sealing washer
6. Rear pin (with IPTO)
7. Front pin (with IPTO)
8. Bushing
9E. Load control shaft (early)
9L. Load control shaft (later)
15. Special pin
16. Special screw
17. Jam nut
18. "O" ring

19. Control arm shaft
20. Load control arm
21. Jam nut
22. Retaining ring
23. Adjusting screw
24. Control arm extension
25. Pin
26. Pin
27. Cam follower
28. Spring

moves load control arm (20), which pivots on shaft (19). Movement of load control arm is transmitted to the rockshaft control valves via cam follower (27) and roller link (2—Fig. 193). The control linkage and control valves are opened or closed by this movement permitting oil to flow to or from the rockshaft cylinder.

180. R&R AND OVERHAUL. To remove the load sensing mechanism, remove rockshaft housing assembly as outlined in paragraph 177 and the three-point hitch. Remove cam follower spring (28—Fig. 194A) and screw ¼ inch pipe plug, located near left final drive housing, out of transmission case. Slide pivot shaft (19) to the left out through pipe plug hole far enough so control arm (20) can be removed. Remove pin, retainer and bushing (4) from right end of load control shaft (9E or 9L) and bump shaft from transmission case. Negative stop screw (16) can also be removed if necessary.

Remove bearing housings with bushings (8). Renew bushings as necessary. New bushings are installed with larger chamfer out toward seal (5). Check special

pin (15) for damage in area where it is contacted by load control shaft and renew if necessary. Also check contact areas of negative stop screw (16) or pins (6 and 7) and load control arm (20). Check load control shaft (9E or 9L) to be sure it is not bent or otherwise damaged. Wear or damage to any other parts will be obvious.

Renew all "O" rings and seal rings and reassemble load sensing assembly by reversing disassembly procedure. When installing hitch draft links, check clearance between bushings and draft links. Add or remove shims at both sides to obtain a clearance of 2.5-3.7 mm (0.100-0.140 inch). After assembly is completed, adjust negative stop screw as outlined in paragraph 170.

REMOTE CONTROL SYSTEM

Tractors may be equipped with single or dual selective (remote) control valves. Early models are equipped with poppet-type valves and later models are usually equipped with spool-type valves. Valves are mounted on a bracket attached to right final drive housing. Each control valve will operate a single or double acting remote cylinder.

SELECTIVE CONTROL VALVE

All Models (Spool Valve)

181. R&R AND OVERHAUL. To remove the valve or valves, disconnect hydraulic lines from valves and on models with Sound Gard Body, disconnect control lever linkage. Remove cap screws (A—Fig. 195A) and remove valves. Remove the four cap screws (B) and separate valves.

To remove valve cover and disassemble a valve set, drive pin (22—Fig. 195) from cover (21) and withdraw pinion shaft (25). Unbolt and remove cover (21), remove pins (19) and remove racks (20). Spool valve bodies (8) have hydraulic passages open to both sides. Valve spools and body are available only as a matched set and should never be interchanged with spool or body from another valve. One valve body (16) in a set has hydraulic openings open to only one side. Closed valve body (16) is assembled with three washers (9 and 18) that are identical. Single acting selective control valves are similar to double acting shown except only one valve spool and one rack is used.

Spring (12) should have a free length of 26 mm (1.02 inches) and should exert 210-260 N (49-58 lbs.) when compressed to 15 mm (0.59 inch). The area of valve spool at "V" seal (10) should be 9.96-10.00 mm (0.388-0.390 inch) and should not show evidence of wear. When assembling, sealing lips of seals (10) should be toward inside and the proper number of washers (9 and 18) should be installed. Tighten screws (24) to 15 N·m (11 ft.-lbs.) torque. Coat pinion shaft (25) with John Deere EP multi-purpose grease, push valves into body until inner washer (11) is against shoulder of valve body and insert pinion shaft (25) so that bore shaft is aligned with bore in cover for plug (23). Drive pin (22) into cover with slot away from pinion shaft. On models with Sound Gard Body, install lever (A—Fig. 195B) pointing downward at 32 degrees as shown. On all models, tighten the four screws (B—Fig. 195A) to 20 N·m (14 ft.-lbs.) and screws (A) to 50 N·m (35 ft.-lbs.) torque.

All Models (Poppet Valve)

182. R&R AND OVERHAUL. To remove the valve or valves, disconnect remote coupling lines, inlet line and return line from valve. On models with Sound Gard Body, disconnect control lever linkage, unbolt valve from mounting brackets and remove valve. On models without Sound Gard Body, unbolt bracket from final drive housing and lift off valve and bracket assembly.

To overhaul the removed valve, refer to Fig. 196 and proceed as follows: Identify and remove control levers, then unbolt and remove end cap (37), which will

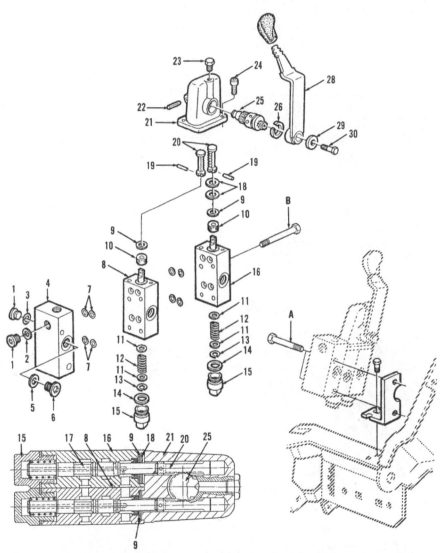

Fig. 195—Exploded and cross section views of spool type selective (remote) control valve. Refer to Fig. 196 for poppet type.

1. Plug
2. Seal ring
3. Seal ring
4. Valve stack end
5. Seal ring
6. Plug
7. "O" rings
8. Spool & housing with through passages
9. Washer
10. Seal
11. Washer
12. Spring
13. Snap ring
14. Seal ring
15. End cap
16. Spool & end housing
17. Spool
18. Washers (same as 9)
19. Pins
20. Racks
21. Cover
22. Spring pin
23. Plug
24. Screws
25. Pinion shaft
26. Seal
28. Handle
29. Washer
30. Retaining screw

Illustration for Fig. 195 reproduced by permission of Deere & Company. Copyright Deere & Company.

contain metering valve (50). Note while loosening, that the four valve guides (33) are spring loaded and retained by the end cap. The four guides should move out with the cap as cap screws are loosened. If they do not, protect them from flying out as cap is removed. Withdraw guides (33) and springs (32 and 55), valves (31) and associated parts, keeping them identified and in proper order.

Note: Pressure springs are located on valves in location (P—Fig. 196A) and should be color-coded red. Return springs, color-coded green, should be on valves located in ports (R).

Remove flow control valve (46—Fig. 196) and spring (47). Remove snap ring (26) and outer guide (23), then withdraw spring (36), detent cartridge assembly (18, 19, 20, 21, 22, 27 and 54), detent follower assembly (15, 16 and 17).

Invert valve body, rocker assembly end up, then drive out spring pin (8) securing rocker (10) to lever shaft (60), withdraw shaft and lift out rocker assembly.

Clean all parts and inspect housing, side cover and end caps for cracks, nicks or burrs. Small imperfections can be removed with a fine file. However, parts should be renewed if their condition is questionable. Inspect poppet valves (31) and their seats in housing (14) for grooves, scoring or excessive wear and renew parts as necessary. Check springs against the values that follow:

Flow control valve spring (47)—
Free length . . . . 47 mm (1.85 inches)
Test at 39 mm
(1.5 inches) . . . . 160-195 N (36-44 lbs.)

Pressure valve springs (32 or 55)—
Color . . . . (red)
Free length . . . . 44 mm (1.72 inches)
Test at 31 mm
(1.2 inches) . . . . 160-195 N (36-44 lbs.)

Return valve springs (32 or 55)—
Color . . . . (green)
Free length . . . . 39 mm (1.5 inches)
Test at 31 mm
(1.2 inches) . . . . 80-100 N (18-22 lbs.)

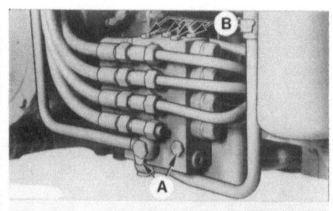

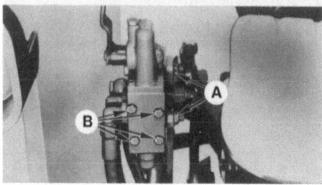

Fig. 195A—Refer to text for proper removal and installation of selective control valves.

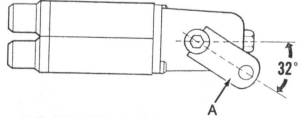

Fig. 195B—Lever (A) should be installed pointing downward at 32 degree angle as shown for models with Sound Gard Body.

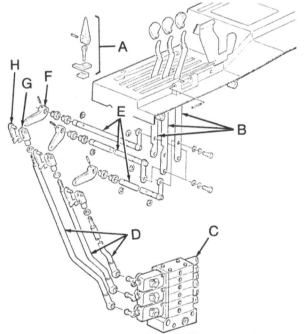

Fig. 195C—Drawing of typical linkage connecting control handles to selective control valves on models with Sound Gard Body.

A. Stop (for hyd. motor operation)
B. Control levers
C. Selective control valves
D. Connecting rods
E. Shafts
F. Lever
G. Fork end
H. Safety strap

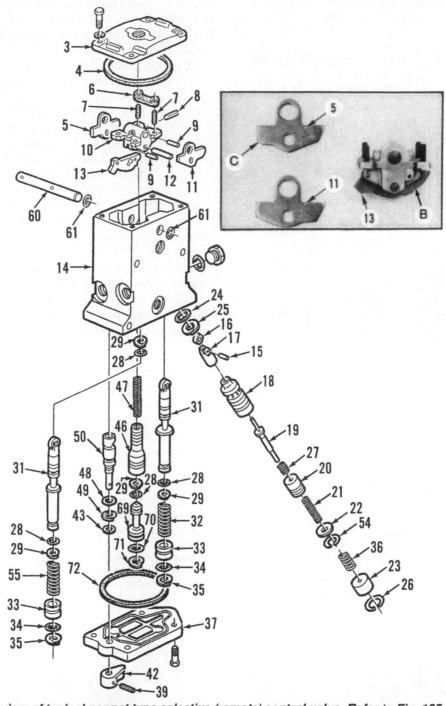

Fig. 196—Exploded view of typical poppet type selective (remote) control valve. Refer to Fig. 195 for spool type valve.

3. Cover	16. Roller	29. Back-up ring	48. "O" ring
4. Packing	17. Detent follower	31. Valves	49. Back-up ring
5. Regular cam	18. Detent cartridge	32. Spring	50. Metering valve
6. Rubber keeper	19. Detent pin	33. Valve guide	54. Snap ring
7. Adjusting screw	20. Detent piston	34. "O" ring	55. Spring
8. Spring pin	21. Spring	35. Back-up ring	60. Shaft
9. Drive pin	22. Special washer	36. Spring	61. "O" ring
10. Rocker	23. Outer guide	37. End cap	69. Flow control
11. Float cam	24. "O" ring	39. Pin	valve stop
12. Pin	25. Back-up ring	42. Metering lever	70. "O" ring
13. Detent cam	26. Snap ring	43. Thrust washer	71. Back-up ring
14. Housing	27. Spring	46. Flow control valve	72. Packing
15. Spring pin	28. "O" ring	47. Spring	

Cam detent follower spring (36)—
Free length . . . . 19 mm (0.75 inch)
Test at 8 mm (0.31 inch) . . . . 58-70 N (13-16 lbs.)

Spring (27)—
Free length . . . . 19.4 mm (0.76 inch)
Test at 14 mm
(0.56 inch) . . . . 89-108 N (20-24.4 lbs.)

Spring (21)—
Free length . . . . .35 mm (1.38 inches)
Test at 26 mm
(1.02 inches) . . . . 51-62 N (11.5-14 lbs.)

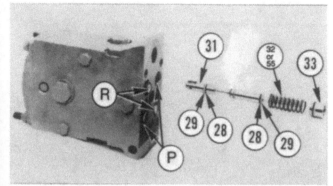

Fig. 196A—Pressure valves are installed in bores (P) with RED springs and return valves are installed in bores (R) with GREEN springs.

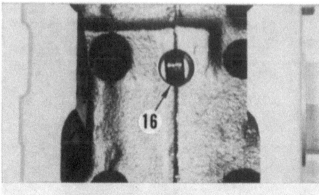

Fig. 197—View of selective control valve with rocker removed and with rocker (10) and shaft (60) being installed. Refer to text when assembling.

Wash housing with soap and water to remove any contamination, then dip all parts except keepers (6) in John Deere Hydraulic and Transmission oil before assembling. Carefully install new "O" ring (24) and back-up ring (25) in housing bore, then install preassembled cartridge (18, 19, 20, 21, 22, 27 and 54), spring (36), guide (23) and snap ring (26). Install new "O" rings (61) in housing bores. Assemble rocker (5, 7, 9, 10, 11, 12 and 13) using Fig. 196 as a guide. The large float stop of detent cam (13) at arrow should be assembled opposite the larger pin boss (B). The regular cam (5) can be identified by chamfer (C) and different shape than float cam (11). Insert detent follower (15, 16 and 17) with roller (16—Fig. 197) aligned with center line of housing, then install rocker assembly with detent cam (13) located as shown. Install control shaft (60) with lever positioned correctly, then install pin (8—Fig. 196) through rocker and shaft. Install new "O" rings (29) and back up rings (28) on valves (31), then install valves (31) in proper bores as shown in Fig. 196A. Adjust valve as follows before installing springs (32 or 55) and guides (33). Make sure that roller on valve is properly aligned with detents (5 and 11).

Selective control valve is adjustable and if disassembled, valve must be adjusted during assembly. Adjustment requires the use of a special adjusting cover JDH-15C and a dial indicator as shown in Fig. 199. With valve guides (33—Fig. 196) removed, install adjusting cover (F—Fig. 199) with washers between housing and cover to ensure cover is level and secure.

NOTE: With selective control valves in neutral, pressure and return valves are adjusted by setting a specified distance between the operating cams and the valve rollers. This specified distance ensures that return valves open before the pressure valves and that maximum oil flow through selective control valves can be maintained.

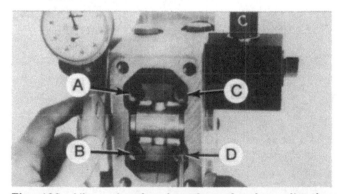

Fig. 198—View showing location of valve adjusting screws.

A. Pressure valve
    adjusting screw
B. Return valve
    adjusting screw
C. Return valve
    adjusting screw
D. Pressure valve
    adjusting screw

Loosen all four screws (A, B, C and D—Fig. 198) to make sure there is some clearance. Finger tighten the four screws (G, H, J and K—Fig. 199) to seat the valves. Finger tighten detent screw (E) while moving control lever to locate neutral position. Install a dial indicator so indicator pin contacts control lever 51 mm (2 inches) from centerline of control shaft and at right angles to lever. Dial indicator should have a minimum travel of 3.55 mm (0.140 inch) in each direction from zero. Set indicator at zero, which is detent cam zero. Turn each of the four adjusting screws (A, B, C and D—Fig. 198) in, one at a time, until dial indicator just begins to move, then back adjusting screws for pressure valves (A and D) out ½ turn and return adjusting screws (B and C) out ¼ turn.

Back out detent screw (E—Fig. 199) and screws (J and K) on one side. Check to make sure screws (G and H) on other side are still tight. At opposite end of valve, adjust pressure screw (C—Fig. 198) so upward movement of control lever will read 0.5-0.8 mm (0.020-0.030 inch) on dial indicator. Turn screw IN to decrease dial indicator reading or OUT to increase. After adjusting screw (C), adjust pressure screw (D) so downward movement of control lever will read 1.5-1.8 mm (0.060-0.070 inch) on dial indicator. Adjust screws (C and D) of valves that are held against seats by screws (G and H—Fig. 199).

Tighten screws (J and K) and loosen screws (G and H). At opposite end of valve, adjust return screw (B—Fig. 198) until downward movement of control lever will read 0.5-0.8 mm (0.020-0.030 inch) from zero. Then, adjust return screw (C) until upward movement of control lever will read 1.5-1.8 mm (0.060-0.070 inch) from zero on dial indicator. Because this adjustment is so critical, it is advisable to adjust a second time using the same sequence.

With adjustment completed, remove adjustment cover (F—Fig. 199). Reassemble valve using new seals and gaskets. Springs color-coded RED should be installed on pressure valves at locations (P—Fig. 196A) and GREEN springs should be installed on return valves (R). Tighten control valve end cap and cover cap screws to a torque of 50 N·m (35 ft.-lbs.).

BREAKAWAY COUPLER

ISO Poppet Coupler

184. To disassemble the ISO poppet-type remote coupler, remove plug (15—Fig. 201 or Fig. 202), screws (14 and 16), then remove dust cover assembly (5 through 8). Remove four screws (13) and separate lever and cover assembly (1) from housing (3). Using special tool JDG266 or equivalent, depress sleeve (10) and remove snap ring (9), sleeve (10), spring (11) and receptacle (12). Remove the 10 balls from receptacle. Two types of receptacles are used.

To disassemble the coupler receptacle of the type shown in Fig. 202, clamp coupling (21) softly in a soft jawed vise and unscrew socket (37). Disassembly, inspection and reassembly procedures will be evident. Always install new "O" rings, back-up rings and other soft parts when assembling. Retaining spring (11) should test 140-170 N (32-39 lbs.) when compressed to 19 mm (0.75 inch). Poppet valve spring (25) should test 75-90 N (16-20 lbs.) when compressed to 16.5 mm (0.64 inch). Bleed valve spring (35) should test 8-10 N (1.8-2.2 lbs.) when compressed to 1.65 mm

Fig. 199—Install dial indicator so indicator pin contacts control lever on a line 51 mm (2 inches) from centerline of control shaft. Install valve adjusting tool JDH-15C as shown.

E. Detent screw   H. Pressure screw
F. Adjusting cover JDH-15C   J. Pressure screw
G. Return screw   K. Return screw

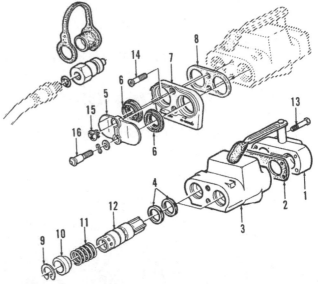

Fig. 201—Exploded view of one type breakaway coupler. Refer to Fig. 202 for other type.

1. Cover	9. Snap ring
2. Gasket	10. Sleeve
3. Housing	11. Spring
4. Seals	12. Receptacle
5. Cover	13. Screws
6. Grommets	14. Screw
7. Retainer	15. Plug
8. Gasket	16. Screw

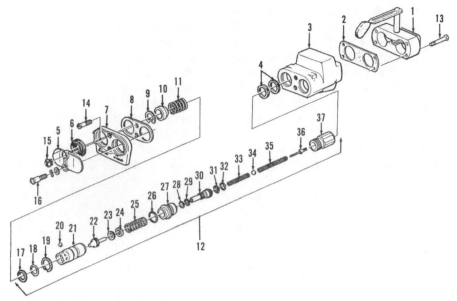

Fig. 202—Exploded view of one type breakaway coupler. Refer to Fig. 201 for other type.

1. Lever & cover assy.			
2. Gasket	11. Spring	20. Balls (10)	29. Back-up ring
3. Housing	12. Receptacle	21. Coupler	30. Piston
4. Seal rings	13. Screws	22. Cone	31. Back-up ring
5. Cover	14. Screw	23. Rubber washer	32. "O" ring
6. Grommet	15. Plug	24. Steel washer	33. Short spring
7. Retainer	16. Screw	25. Spring	34. Ball
8. Gasket	17. Back-up ring	26. "O" ring	35. Longer spring
9. Snap ring	18. "O" ring	27. Guide	36. Bleed valve
10. Sleeve	19. Snap ring	28. "O" ring	37. Coupler socket

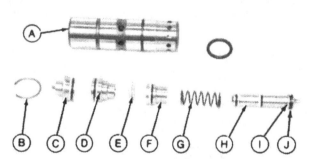

Fig. 203—Exploded view of ISO coupler receptacle.

A. Coupler receptacle	F. Rear piston
B. Snap ring	G. Spring
C. Bleed valve	H. Poppet
D. Front piston	I. Spacer
E. Snap ring	J. Packing

Fig. 204—Exploded view of bleed valve assembly (C—Fig. 203).

A. Bleed valve
B. Spring
C. Snap ring

(0.64 inch). Tighten coupler socket (37) to 50 N·m (35 ft.-lbs.) torque. Renew all "O" rings, back-up rings, seal rings and gaskets and reassemble by reversing the disassembly procedure. Tighten cap screws(13) retaining release lever and cover (1) to housing (3) to a torque of 8 N·m (6 ft.-lbs.). Tighten screws attaching coupler housing to bracket to 50 N·m (35 ft.-lbs.) torque. When installing breakaway coupler bracket to rockshaft housing, tighten cap screws to a torque of 120 N·m (85 ft.-lbs.).

To disassemble coupler receptacle of the type shown in Figs. 203, Fig. 204 and Fig. 205, proceed as follows. Remove snap ring (B—Fig. 203), bleed valve (C) and front piston (D). Remove snap ring (E), then withdraw rear piston (F), spring (G) and poppet (H). To disassemble bleed valve (C), refer to Fig. 204, remove snap ring (C), then remove spring (B) and bleed valve poppet (A).

Clean and inspect all parts. Spring (11—Fig. 201) should test 140-170 N (32-39 lbs.) when compressed to a length of 19 mm (0.75 inches). Spring (G—Fig. 203) should test 75-90 N (16-20 lbs.) when compressed to a length of 16.5 mm (0.64 inch). Spring (B—Fig. 204) should test 8-10 N (1.8-2.2 lbs.) when compressed to a length of 5 mm (0.20 inch). Renew all "O" rings, back-up rings, seal rings and gaskets and reassemble by reversing the disassembly procedure. Tighten cap screws(13—Fig. 201) retaining release lever and cover (1) to housing (3) to a torque of 8 N·m (6 ft.-lbs.). When installing breakaway coupler on tractor, tighten cap screws to a torque of 120 N·m (85 ft.-lbs.).

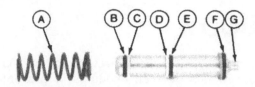

Fig. 205—View showing placement of "O" rings and back-up rings on poppet.

A. Spring
B. "O" ring
C. Back-up ring
D. Back-up ring

E. "O" ring
F. Packing
G. Poppet

REMOTE CYLINDER

185. To disassemble the double-acting remote cylinder, remove oil lines and end cap (18—Fig. 206). Remove stop valve (14) by pushing stop rod (9) completely into cylinder. Remove nut (21) from piston rod, then remove piston (24) and rod (34). Push stop rod (9) all the way into cylinder and drift out pin (27). Remove piston rod guide (26).

Renew all seals and examine other parts for wear or damage. Wiper seal (35) should be installed with lip toward outer end of bore. Install stop rod seal assembly (1, 2 and 3) with sealing edge toward cylinder. Complete the assembly by reversing the disassembly procedure. Tighten end cap screws to 160 N·m

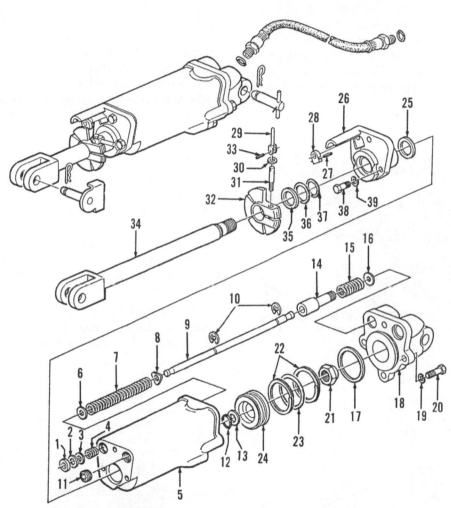

Fig. 206—Exploded view of typical remote cylinder.

1. Adapter	11. Plug	21. Nut
2. Packing	12. "O" ring	22. Back-up ring
3. Washer	13. Back-up ring	23. "O" ring
4. Spring	14. Stop valve	24. Piston
5. Cylinder	15. Spring	25. Gasket
6. Washer	16. Gasket	26. Guide
7. Spring	17. Gasket	27. Pin
8. Washer	18. Cap	28. Stop rod arm
9. Stop rod	19. Lockwasher	29. Stop lever
10. Snap ring	20. Cap screw	30. Washer

31. Stop screw	
32. Rod stop	
33. Pin	
34. Piston rod	
35. Seal	
36. Washer	
37. "O" ring	
38. Cap screw	
39. Lockwasher	

(120 ft.-lbs.) torque and piston rod guide screws to 50 N•m (35 ft.-lbs.) torque.

To adjust the working stroke, lift piston stop lever (29), slide adjustable stop (32) along piston rod to the desired position and press the stop lever down. If clamp does not hold securely, lift and rotate stop lever ½-turn clockwise and reset. Make certain that adjustable stop is located so that the stop rod contacts one of the flanges on adjustable stop.

SOUND GARD BODY

Models So Equipped

186. REMOVE AND REINSTALL. To remove the Sound Gard Body, unbolt and remove left and right platform steps. Disconnect battery cables, then remove batteries and left hand battery box. Drain coolant, remove radiator side grilles side panels and hood. Disconnect engine wiring harness at dash and disconnect steering hoses at connectors behind dash. Immediately plug line openings. Disconnect shut-off cable and speed control rod at fuel injection pump, then snap rear of speed control rod out of ball joint and remove rod. Remove rate-of-drop adjusting screw, left and right dash panels and the center platform plate.

Fig. 210—Check to be sure that all control rods (A), hoses and connectors (B) are correctly positioned before lowering cab onto tractor.

Drain systems, then disconnect brake and clutch hydraulic lines. Cover all openings to prevent the entrance of dirt into systems. Disconnect Hi-Lo shift rod, pto control rod, electrical connector plug and return line from top of transmission shift cover. Unplug rear electrical harness connector. If so equipped, disconnect air conditioning lines at couplers. Disconnect and lower the auxiliary fuel tank. Disconnect differential lock rod and after noting length of hand brake rod, disconnect brake rod from rear yoke. Disconnect left and right brake hoses and cap or plug openings. Fully lower rockshaft lift arms, then detach lift link adjusters from cab frame. Remove rockshaft and selective control valve linkage rods. **Remove rods completely to prevent damage.** Loosen hose clamps and disconnect heater hoses. Disconnect transmission shift linkage and check for any hoses from cab that may be secured to parts that will remain with tractor. Unbolt Sound Gard Body supports, noting the arrangement of the support pads and washers.

Various methods of removing the Sound Gard Body may be used. When using a lifting bar, remove the two cap screws at top of Sound Gard Body and install two lifting eyebolts. Attach lifting crossbar and overhead hoist to the eyebolts. Carefully lift Sound Gard Body from tractor.

When installing Sound Gard Body, reverse removal procedure. Be sure rod and hoses are properly positioned before lowering body into position. Check for

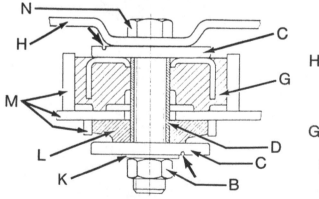

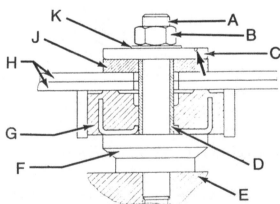

Fig. 212—Sound Gard Body mounts should be assembled as shown. Marked side of washers (C) indicated by arrows should be away from rubber parts.

A. Stud	E. Final drive housing	H. Body frame	L. Rubber washer
B. Nut	F. Washer	J. Rubber washer	M. Mounting bracket
C. Washers	G. Rubber bearing pad	K. Washer	N. Cap screw
D. Bushing			

correct placement of support mounting pads and washers as shown in Fig. 212. Tighten support nuts to a torque of 200 N·m (145 ft.-lbs.). Bleed steering, brake and clutch systems as described in appropriate paragraphs. Clean and apply refrigerant oil to air conditioning line connectors before attaching, then check air conditioning charge as required.

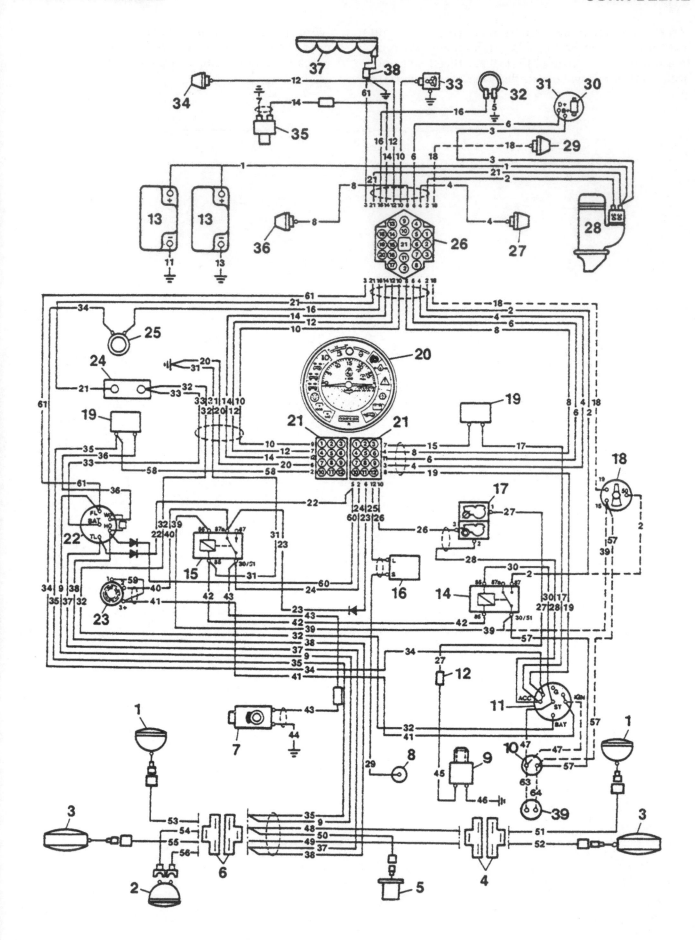

Illustration for Fig. 220 reproduced by permission
of Deere & Company. Copyright Deere & Company.

Fig. 220—Wiring diagram typical of models without Sound Gard Body.

Components
1. Headlights
2. Work lights/Tail lights
3. Warning lights
4. Rt. connector
5. Socket for headlamp
6. Lf. connector
7. Handbrake control light sender
8. Transmission oil pressure warning switch
9. Solenoid for MFWD
10. Neutral start switch
11. Main switch
12. Connector
13. Batteries
14. Relay for handbrake control light (w/o Thermostart starting aid)
15. Relay for handbrake control light
16. Time delay switch for transmission oil pressure
17. MFWD control switch
18. Start switch (with Thermostart starting aid)
19. Flasher
20. Instrument cluster
21. Plug
22. Light switch
23. Handbrake warning horn
24. Circuit breaker
25. Horn button
26. Plug
27. Engine oil pressure warning switch
28. Starting motor
29. Burner of Thermostart starting aid
30. Regulator
31. Alternator
32. Horn

33. Fuel gage sending unit
34. Air cleaner restriction warning switch
35. Magnetic sender for Speed-Hour meter
36. Coolant temperature sending unit
37. Light bar
38. Connector
39. Neutral start switch (with continuous running pto)

Wires
1. Black (2 Ga.)
2. Black (14 Ga.)
3. Red (12 Ga.)
4. Light blue/green (16 Ga.)
5. Brown (16 Ga.)
6. Blue (16 Ga.)
7. Brown (16 Ga.)
8. Green/yellow (16 Ga.)
9. Blue (16 Ga.)
10. Light blue/black (16 Ga.)
11. Ground strap
12. Black/white (16 Ga.)
13. Ground strap
14. Gray/red (16 Ga.)
15. Black/white/green (16 Ga.)
16. Black/yellow (16 Ga.)
17. Black/white/green (16 Ga.)
18. Gray/black (14 Ga.)
19. Black (16 Ga.)
20. Black/white/green (16 Ga.)
21. Red (12 Ga.)
22. Gray (16 Ga.)
23. Red (16 Ga.)
24. Black/red (16 Ga.)
25. Green/white (16 Ga.)
26. Green (16 Ga.)
27. Black/yellow (16 Ga.)

28. Green (16 Ga.)
29. Green/white (16 Ga.)
30. Black (16 Ga.)
31. Brown (16 Ga.)
32. Red (16 Ga.)
33. Red (16 Ga.)
34. Black/yellow (16 Ga.)
35. Orange (16 Ga.)
36. Orange/white (16 Ga.)
37. Yellow (16 Ga.)
38. Pink (16 Ga.)
39. Brown (16 Ga.)
40. Red (16 Ga.)
41. Black (16 Ga.)
42. Brown (16 Ga.)
43. Black/yellow/red (16 Ga.)
44. Brown (16 Ga.)
45. Green/yellow (16 Ga.)
46. Brown (16 Ga.)
47. Black (14 Ga.)
48. Pink (16 Ga.)
49. Orange (16 Ga.)
50. Yellow (16 Ga.)
51. Pink (16 Ga.)
52. Orange (16 Ga.)
53. Pink (16 Ga.)
54. Yellow (16 Ga.)
55. Orange (16 Ga.)
56. Blue (16 Ga.)
57. Black (14 Ga.)
58. Orange (16 Ga.)
59. Black/red (16 Ga.)
60. Gray (16 Ga.)
61. Blue/white (16 Ga.)
62. Blue/white (16 Ga.)
63. Black (14 Ga.)
64. Black (14 Ga.)

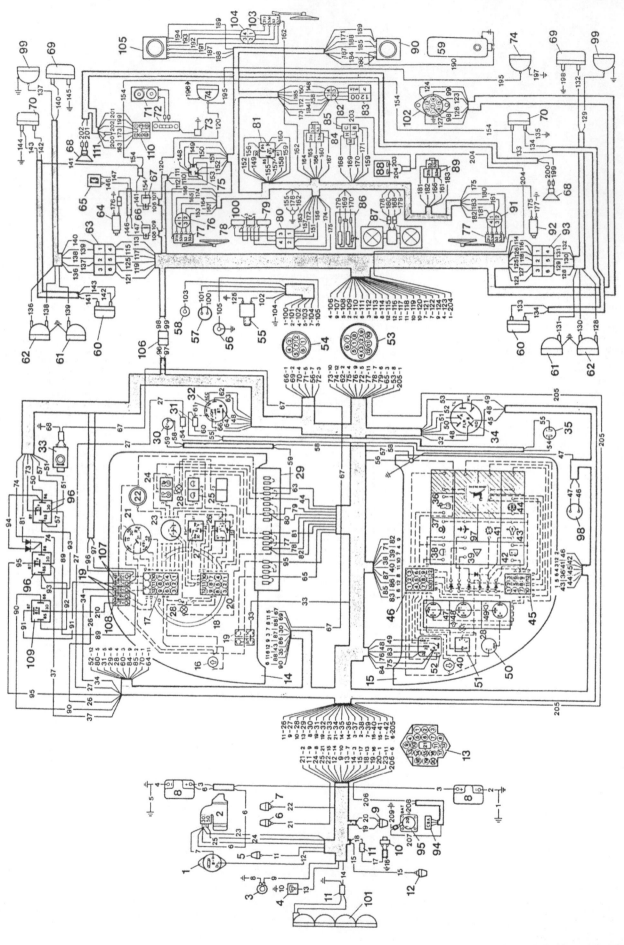

Illustration for Fig. 221 reproduced by permission
of Deere & Company. Copyright Deere & Company.

Fig. 221—Wiring diagram typical of early models (before SN 491028L) with Sound Gard Body.

Components

1. Alternator & regulator
2. Starting motor
3. Horn
4. Fuel gage sending unit
5. Solenoid for ether starting aid or burner for Thermostart aid
6. Oil pressure gage sending unit
7. Oil pressure indicator light sending unit
8. Batteries
9. Coolant temp. sending unit
10. Eng. & tractor speed magnetic sending unit
11. Connector
12. Air cleaner restriction warning switch
13. Plug
14. Rt. instrument housing
15. Lf. instrument housing
16. Rt. turn signal indicator light
17. Plug
18. Speed-hour meter
19. Circuit breaker (30 amp.)
20. Plug
21. Hazard warning light switch
22. Horn button
23. Handbrake warning horn
24. MFWD switch
25. High-Low beam headlight switch
26. Starter relay
27. Handbrake indicator light relay
28. Instrument lighting
29. Circuit board
30. Socket for headlamp
31. Connector
32. Main switch
33. Handbrake indicator sending unit
34. Light switch
35. Button for ether starting aid
36. MFWD indicator light
37. Handbrake indicator light
38. Headlight high beam indicator light
39. Warning lights
40. Lf. Turn signal light
41. Transmission oil pressure indicator light
42. Alternator indicator light
43. Engine oil pressure indicator light
44. Air cleaner restriction indicator light
45. Plug
46. Plug
47. Fuel gage
48. Engine oil temperature gage
49. Coolant temperature gage
50. Flasher
51. Time delay switch for transmission oil pressure
52. Turn signal switch
53. Plug
54. Plug
55. MFWD solenoid
56. Hi-Lo shift unit control light sending unit
57. Neutral start switch
58. Transmission oil pressure warning switch
59. Windshield washer system
60. Front turn signal & warning lights
61. Front work light
62. Headlights
63. Plug
64. Rt. door switch for cab light
65. Hi-Lo shift unit control light
66. Plug
67. Plug
68. Loud speaker
69. Taillights
70. Rear turn signal & warning lights
71. Tape player
72. Radio
73. Electronic aerial
74. Upper rear worklight
75. Plug
76. Plug
77. Front windshield wiper
78. Gear shift console light
79. Cab interior light
80. Plug
81. Fan relay
82. Plug
83. Clock
84. Switch for Rt. windshield wiper
85. Switch for cab heater & blower fan
86. Resistors
87. Cab heater & blower fan
88. Thermostat switch
89. Switch for Lf. windshield wiper
90. Windshield washer switch
91. Plug
92. Lf. door switch for cab light
93. Plug
94. Thermal fuse (air conditioning)
95. AC compressor
96. Worklight relay
97. Clutch indicator light
98. Clutch indicator light sending unit
99. Lower rear worklights
100. Cab interior light switch
101. Light bar
102. 7-Terminal socket
103. Rear windshield wiper
104. Plug
105. Rear windshield wiper switch
106. Connector
107. Circuit breaker (20 amp.)
108. Circuit breaker (10 amp.)
109. Relay for 7-terminal socket
110. Radio plug
111. Fuse (5 amp.)

Wires

1. Black (2 Ga.)
2. Ground strap
3. Black (2 Ga.)
4. Ground strap
5. Black (2 Ga.)
6. Black (1 Ga.)
7. Red (12 Ga.)
8. Brown (16 Ga.)
9. Black/yellow (16 Ga.)
10. Brown (16 Ga.)
11. Gray/black (14 Ga.)
12. Blue (16 Ga.)
13. Blue/white (16 Ga.)
14. Blue/white (16 Ga.)
15. White/black (16 Ga.)
16. Brown (16 Ga.)
17. Green/red (16 Ga.)
18. Gray/red (16 Ga.)
19. Green/yellow (16 Ga.)
20. Black/violet (16 Ga.)
21. Green (16 Ga.)
22. Blue/green (16 Ga.)
23. Red (11 Ga.)
24. Black (14 Ga.)
25. Red (12 Ga.)
26. Red (11 Ga.)
27. Gray/black (14 Ga.)
28. Black/yellow (16 Ga.)
29. Gray/red (14 Ga.)
30. Black (14 Ga.)
31. Green/black (16 Ga.)
32. Black (16 Ga.)
33. Red (11 Ga.)
34. Red (14 Ga.)
35. Blue (16 Ga.)
36. Blue (16 Ga.)
37. Blue/white (16 Ga.)
38. Green (16 Ga.)
39. Blue/black (16 Ga.)
40. Green/yellow (16 Ga.)
41. Blue/green (16 Ga.)
42. White/black (16 Ga.)
43. Red (16 Ga.)
44. Black (16 Ga.)
45. White (16 Ga.)
46. Black/violet (16 Ga.)
47. Brown (16 Ga.)
48. Red (16 Ga.)
49. Red/yellow (16 Ga.)
50. Blue (16 Ga.)
51. Red (16 Ga.)
52. Red (16 Ga.)
53. Gray/black (16 Ga.)
54. Gray/black (14 Ga.)
55. Gray/black (14 Ga.)
56. Brown (11 Ga.)
57. Brown (16 Ga.)
58. Brown (16 Ga.)
59. Black (16 Ga.)
60. Black/violet (14 Ga.)
61. Black/violet (16 Ga.)
62. Black (15 Ga.)
63. Black/yellow (16 Ga.)
64. Red (14 Ga.)
65. Red. (14 Ga.)
66. Black (14 Ga.)
67. Black/pink (16 Ga.)
68. Brown (16 Ga.)
69. Black (16 Ga.)
70. Green/yellow (16 Ga.)
71. Green/white (16 Ga.)
72. Brown/white (16 Ga.)
73. Blue/green (14 Ga.)
74. Blue/white (16 Ga.)
75. Orange/black (16 Ga.)
76. Orange (16 Ga.)
77. Beige (16 Ga.)
78. Green/white (16 Ga.)
79. Gray (16 Ga.)
80. White (16 Ga.)
81. Green (16 Ga.)
82. White (16 Ga.)
83. Orange (16 Ga.)
84. Orange/black (16 Ga.)
85. Gray/green (16 Ga.)
86. Black/pink (16 Ga.)
87. Brown (16 Ga.)
88. Black (16 Ga.)
89. Red (14 Ga.)
90. Blue (16 Ga.)
91. Red (11 Ga.)
92. Brown (16 Ga.)
93. Brown (16 Ga.)
94. Blue/green (15 Ga.)
95. Red (15 Ga.)
96. Blue (14 Ga.)
97. Red (14 Ga.)
98. Blue (14 Ga.)
99. Red (14 Ga.)
100. Black (14 Ga.)
101. Black (14 Ga.)
102. Green/yellow (16 Ga.)
103. Green/white (16 Ga.)
104. Brown (11 Ga.)
105. Brown/white (16 Ga.)
106. Orange/black (16 Ga.)
107. Orange (16 Ga.)
108. Brown/white (16 Ga.)
109. Black (15 Ga.)
110. Black (16 Ga.)
111. Gray (16 Ga.)
112. Red (14 Ga.)
113. Gray/black (16 Ga.)
114. Gray/black (16 Ga.)
115. Blue/white (16 Ga.)
116. Blue/white (16 Ga.)
117. Beige (16 Ga.)
118. Beige (16 Ga.)
119. Beige/green (14 Ga.)
120. Blue/green (16 Ga.)
121. Green (16 Ga.)
122. Green (16 Ga.)
123. Orange/black (16 Ga.)
124. Orange (16 Ga.)
125. Blue/green (14 Ga.)
126. Yellow (16 Ga.)
127. Black (15 Ga.)
128. Green (16 Ga.)
129. Blue/green (16 Ga.)
130. Beige (16 Ga.)
131. Blue/white (16 Ga.)
132. Gray/black (16 Ga.)
133. Orange/black (16 Ga.)
134. Brown (16 Ga.)
135. Brown (16 Ga.)
136. Green (16 Ga.)
137. Blue/green (16 Ga.)
138. Beige (16 Ga.)
139. Blue/white (16 Ga.)
140. Gray/black (16 Ga.)
141. Orange/black (16 Ga.)
142. Orange/black (16 Ga.)
143. Brown (16 Ga.)
144. Brown (16 Ga.)
145. Brown (16 Ga.)
146. Black (16 Ga.)
147. Brown/black (16 Ga.)
148. Red (16 Ga.)
149. Red (14 Ga.)
150. Gray (16 Ga.)
151. Gray (16 Ga.)
152. Black (15 Ga.)
153. Black (16 Ga.)
154. Orange (16 Ga.)
155. Brown (16 Ga.)
156. Black/red (16 Ga.)
157. Brown (16 Ga.)
158. Black (16 Ga.)
159. Black (14 Ga.)
160. Black (16 Ga.)
161. Black (16 Ga.)
162. Brown (16 Ga.)
163. Brown (16 Ga.)
164. Red (16 Ga.)
165. Blue (16 Ga.)
166. Black (16 Ga.)
167. Green (16 Ga.)
168. Yellow (14 Ga.)
169. Red (14 Ga.)
170. Gray (14 Ga.)
171. Gray/black (16 Ga.)
172. Brown (16 Ga.)
173. Black/red (16 Ga.)
174. White (16 Ga.)
175. White (16 Ga.)
176. Brown (16 Ga.)
177. Brown (16 Ga.)
178. Brown (14 Ga.)
179. Yellow (14 Ga.)
180. Brown (16 Ga.)
181. Red (16 Ga.)
182. Blue (16 Ga.)
183. Green (16 Ga.)
184. Brown (16 Ga.)
185. Gray (16 Ga.)
186. Brown (16 Ga.)
187. Brown (16 Ga.)
188. Gray (16 Ga.)
189. Black (16 Ga.)
190. Gray/white (16 Ga.)
191. Black (16 Ga.)
192. Green (16 Ga.)
193. Red (16 Ga.)
194. Blue (16 Ga.)
195. Blue/green (16 Ga.)
196. Brown (16 Ga.)
197. Brown (16 Ga.)
198. Brown (16 Ga.)
199. Black/violet (16 Ga.)
200. Blue (16 Ga.)
201. White (16 Ga.)
202. Green (16 Ga.)
203. Gray/black (16 Ga.)
204. Yellow (16 Ga.)
205. Yellow (14 Ga.)
206. Yellow (14 Ga.)
207. Black (16 Ga.)
208. Black (16 Ga.)
209. Black (16 Ga.)
210. Red. (14 Ga.)

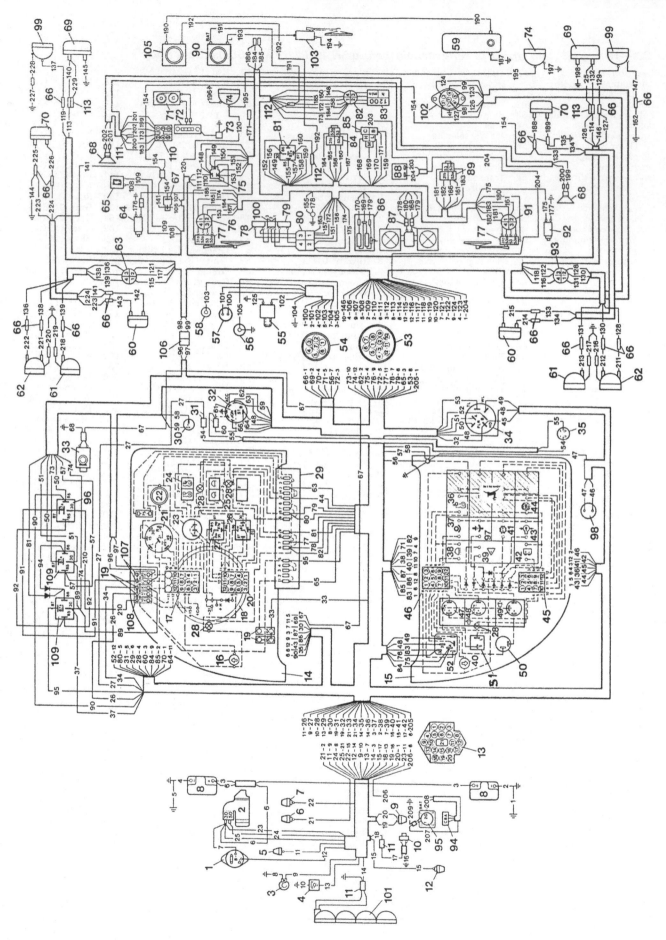

Fig. 222—Wiring diagram typical of later (after SN 491027L) models with Sound Gard Body.

Components
1. Alternator & regulator
2. Starting motor
3. Horn
4. Fuel gage sending unit
5. Solenoid for ether starting aid or burner for Thermostat aid
6. Engine oil pressure gage sending unit
7. Engine oil pressure indicator light sending unit
8. Batteries
9. Coolant temp. sending unit
10. Eng. & tractor speed magnetic sending unit
11. Connector
12. Air cleaner restriction warning switch
13. Plug
14. Rt. instrument housing
15. Lf. instrument housing
16. Rt. turn signal indicator light
17. Plug
18. Speed-hour meter
19. Circuit breaker (30 amp.)
20. Plug
21. Hazard warning light switch
22. Horn button
23. Handbrake warning horn
24. MFWD switch
25. High-Low beam headlight switch
26. Starter relay
27. Handbrake indicator light relay
28. Instrument lighting
29. Circuit board
30. Socket for headlamp
31. Connector
32. Main switch
33. Handbrake indicator sending unit
34. Light switch
35. Button for ether starting aid
36. MFWD indicator light
37. Handbrake indicator light
38. Headlight high beam indicator light
39. Warning lights
40. Lf. Turn signal light
41. Transmission oil pressure indicator light
42. Alternator indicator light
43. Engine oil pressure indicator light
44. Air cleaner restriction indicator light
45. Plug
46. Plug
47. Fuel gage
48. Engine oil temperature gage
49. Coolant temperature gage
50. Flasher
51. Time delay switch for transmission oil pressure
52. Turn signal switch
53. Plug
54. Plug
55. MFWD solenoid
56. Hi-Lo shift unit control light sending unit
57. Neutral start switch
58. Transmission oil pressure warning switch
59. Windshield washer system
60. Front turn signal & warning lights
61. Front work light
62. Headlights
63. Plug
64. Rt. door switch for cab light
65. Hi-Lo shift unit control light
66. Plug
67. Plug
68. Loud speaker
69. Taillights
70. Rear turn signal & warning lights
71. Tape player
72. Radio
73. Electronic aerial
74. Upper rear worklight
75. Plug
76. Plug
77. Front windshield wiper
78. Gear shift console light
79. Cab interior light
80. Plug
81. Fan relay
82. Plug
83. Clock
84. Switch for Rt windshield wiper
85. Switch for cab heater & blower fan
86. Resistors
87. Cab heater & blower fan
88. Thermostat switch
89. Switch for Lf. windshield wiper
90. Windshield washer switch
91. Plug
92. Lf. door switch for cab light
93. Plug
94. Thermal fuse (air conditioning)
95. AC compressor
96. Worklight relay
97. Clutch indicator light
98. Clutch indicator light sending unit
99. Lower rear worklights
100. Cab interior light switch
101. Light bar
102. 7-Terminal socket
103. Rear windshield wiper
104. Plug
105. Rear windshield wiper switch
106. Connector
107. Circuit breaker (20 amp.)
108. Circuit breaker (10 amp.)
109. Relay for 7-terminal socket
110. Radio plug
111. Fuse (5 amp.)
112. Lead connectors
113. Plug

Wires
1. Black (2 Ga.)
2. Ground strap
3. Black (2 Ga.)
4. Ground strap
5. Black (2 Ga.)
6. Black (1 Ga.)
7. Red (12 Ga.)
8. Brown (16 Ga.)
9. Black/yellow (16 Ga.)
10. Brown (16 Ga.)
11. Gray/black (14 Ga.)
12. Blue (16 Ga.)
13. Blue/white (16 Ga.)
14. Blue/white (16 Ga.)
15. White/black (16 Ga.)
16. Brown (16 Ga.)
17. Green/red (16 Ga.)
18. Gray/red (16 Ga.)
19. Green/yellow (16 Ga.)
20. Black/violet (16 Ga.)
21. Green (16 Ga.)
22. Blue/green (16 Ga.)
23. Red (11 Ga.)
24. Black (14 Ga.)
25. Red (12 Ga.)
26. Red (11 Ga.)
27. Gray/black (14 Ga.)
28. Black/yellow (16 Ga.)
29. Gray/red (14 Ga.)
30. Black (14 Ga.)
31. Green/white (16 Ga.)
32. White (16 Ga.)
33. Red (14 Ga.)
34. Red (11 Ga.)
35. Blue (16 Ga.)
36. Blue (16 Ga.)
37. Blue/white (16 Ga.)
38. Green (16 Ga.)
39. Blue/black (16 Ga.)
40. Green/yellow (16 Ga.)
41. Blue/green (16 Ga.)
42. White/black (16 Ga.)
43. Red (16 Ga.)
44. Black (16 Ga.)
45. White (16 Ga.)
46. Black/violet (16 Ga.)
47. Brown (16 Ga.)
48. Red (16 Ga.)
49. Red/yellow (16 Ga.)
50. Blue (16 Ga.)
51. Red (16 Ga.)
52. Red (16 Ga.)
53. Gray/black (16 Ga.)
54. Gray/black (14 Ga.)
55. Gray/black (14 Ga.)
56. Brown (11 Ga.)
57. Brown (16 Ga.)
58. Brown (16 Ga.)
59. Black (16 Ga.)
60. Black/violet (14 Ga.)
61. Black/violet (16 Ga.)
62. Black (15 Ga.)
63. Black/yellow (16 Ga.)
64. Red (14 Ga.)
65. Red. (14 Ga.)
66. Black (14 Ga.)
67. Black/pink (16 Ga.)
68. Brown (16 Ga.)
69. Black (16 Ga.)
70. Green/yellow (16 Ga.)
71. Green/white (16 Ga.)
72. Brown/white (16 Ga.)
73. Blue/green (14 Ga.)
74. Blue/white (16 Ga.)
75. Orange/black (16 Ga.)
76. Orange (16 Ga.)
77. Beige (16 Ga.)
78. Green/white (16 Ga.)
79. Gray (16 Ga.)
80. White (16 Ga.)
81. Green (16 Ga.)
82. White (16 Ga.)
83. Orange (16 Ga.)
84. Orange/black (16 Ga.)
85. Gray/green (16 Ga.)
86. Black/pink (16 Ga.)
87. Brown (16 Ga.)
88. Black (16 Ga.)
89. Red/yellow (14 Ga.)
90. Blue (16 Ga.)
91. Red (11 Ga.)
92. Brown (16 Ga.)
93. Brown (16 Ga.)
94. Blue (16 Ga.)
95. Red (15 Ga.)
96. Blue (14 Ga.)
97. Red (14 Ga.)
98. Blue (14 Ga.)
99. Red (14 Ga.)
100. Black (14 Ga.)
101. Black (14 Ga.)
102. Green/yellow (16 Ga.)
103. Green/white (16 Ga.)
104. Brown (11 Ga.)
105. Brown/white (16 Ga.)
106. Orange/black (16 Ga.)
107. Orange (16 Ga.)
108. Brown/white (16 Ga.)
109. Black (15 Ga.)
110. Black (16 Ga.)
111. Gray (16 Ga.)
112. Red (14 Ga.)
113. Gray/black (16 Ga.)
114. Gray/black (16 Ga.)
115. Blue/white (16 Ga.)
116. Blue/white (16 Ga.)
117. Beige (16 Ga.)
118. Beige (16 Ga.)
119. Blue/green (16 Ga.)
120. Blue/green (16 Ga.)
121. Green (16 Ga.)
122. Green (16 Ga.)
123. Orange/black (16 Ga.)
124. Orange (16 Ga.)
125. Red (16 Ga.)
126. Yellow (16 Ga.)
127. Black (15 Ga.)
128. Green (16 Ga.)
129. Blue (16 Ga.)
130. Beige (16 Ga.)
131. Blue/white (16 Ga.)
132. Yellow (16 Ga.)
133. Orange (16 Ga.)
134. Brown (16 Ga.)
135. Brown (16 Ga.)
136. Green (16 Ga.)
137. Blue (16 Ga.)
138. Beige (16 Ga.)
139. Blue/white (16 Ga.)
140. Yellow (16 Ga.)
141. Orange/black (16 Ga.)
142. Red (15 Ga.)
143. Black (15 Ga.)
144. Brown (16 Ga.)
145. Brown (16 Ga.)
146. Blue/green (14 Ga.)
147. Black (16 Ga.)
148. Red (16 Ga.)
149. Red (14 Ga.)
150. Gray (16 Ga.)
151. Gray (16 Ga.)
152. Black (15 Ga.)
153. Black (16 Ga.)
154. Orange (16 Ga.)
155. Brown (16 Ga.)
156. Black/red (16 Ga.)
157. Brown (16 Ga.)
158. Black (16 Ga.)
159. Black (14 Ga.)
160. Black (16 Ga.)
161. Black (16 Ga.)
162. Brown (16 Ga.)
163. Brown (16 Ga.)
164. Red (16 Ga.)
165. Blue (16 Ga.)
166. Black (16 Ga.)
167. Green (16 Ga.)
168. Yellow (14 Ga.)
169. Red (14 Ga.)
170. Gray (14 Ga.)
171. Gray/black (16 Ga.)
172. Brown (16 Ga.)
173. Black/red (16 Ga.)
174. White (16 Ga.)
175. White (16 Ga.)
176. Brown (16 Ga.)
177. Brown (16 Ga.)
178. Brown (14 Ga.)
179. Yellow (14 Ga.)
180. Brown (16 Ga.)
181. Red (16 Ga.)
182. Blue (16 Ga.)
183. Green (16 Ga.)
184. Brown (16 Ga.)
185. Gray (16 Ga.)
186. Brown (16 Ga.)
187. Brown (16 Ga.)
188. Red (15 Ga.)
189. Black (15 Ga.)
190. Blue (16 Ga.)
191. Red (16 Ga.)
192. Orange (16 Ga.)
193. Yellow (16 Ga.)
194. Black (16 Ga.)
195. Blue/green (16 Ga.)
196. Brown (16 Ga.)
197. Brown (16 Ga.)
198. Black (16 Ga.)
199. Black/violet (16 Ga.)
200. Blue (16 Ga.)
201. White (16 Ga.)
202. Green (16 Ga.)
203. Gray/black (16 Ga.)
204. Yellow (16 Ga.)
205. Yellow (14 Ga.)
206. Yellow (14 Ga.)
207. Black (16 Ga.)
208. Black (16 Ga.)
209. Black (16 Ga.)
210. Red. (14 Ga.)
211. Green (16 Ga.)
212. Tan (16 Ga.)
213. Blue (16 Ga.)
214. Black (15 Ga.)
215. Red (15 Ga.)
216. Black (15 Ga.)
217. Black (16 Ga.)
218. Blue (16 Ga.)
219. Black (16 Ga.)
220. Black (16 Ga.)
221. Tan (16 Ga.)
222. Green (16 Ga.)
223. Brown (16 Ga.)
224. Orange/black (16 Ga.)
225. Black (15 Ga.)
226. Red (15 Ga.)
227. Brown (16 Ga.)
228. Black (16 Ga.)
229. Red (16 Ga.)

Technical Information

Technical information is available from John Deere. Some of this information is available in electronic as well as printed form. Order from your John Deere dealer or call **1-800-522-7448**. Please have available the model number, serial number, and name of the product.

Available information includes:

- PARTS CATALOGS list service parts available for your machine with exploded view illustrations to help you identify the correct parts. It is also useful in assembling and disassembling.
- OPERATOR'S MANUALS providing safety, operating, maintenance, and service information. These manuals and safety signs on your machine may also be available in other languages.
- OPERATOR'S VIDEO TAPES showing highlights of safety, operating, maintenance, and service information. These tapes may be available in multiple languages and formats.

- TECHNICAL MANUALS outlining service information for your machine. Included are specifications, illustrated assembly and disassembly procedures, hydraulic oil flow diagrams, and wiring diagrams. Some products have separate manuals for repair and diagnostic information. Some components, such as engines, are available in separate component technical manuals

- FUNDAMENTAL MANUALS detailing basic information regardless of manufacturer:
  - Agricultural Primer series covers technology in farming and ranching, featuring subjects like computers, the Internet, and precision farming.
  - Farm Business Management series examines "real-world" problems and offers practical solutions in the areas of marketing, financing, equipment selection, and compliance.
  - Fundamentals of Services manuals show you how to repair and maintain off-road equipment.
  - Fundamentals of Machine Operation manuals explain machine capacities and adjustments, how to improve machine performance, and how to eliminate unnecessary field operations.

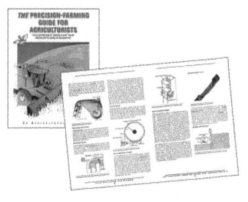

NOTES

NOTES